My Name is Not Susan:

A Love Story Between
Mathematics and Non-Mathematics

By Luke Wolcott

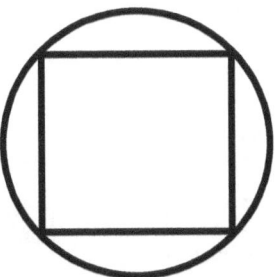

ISBN: 978-0-557-06652-0

"For a long time Mr. Palomar made an effort to achieve such impassiveness and detachment that what counted was only the serene harmony of the lines of the pattern: all the lacerations and contortions and compressions that human reality has to undergo to conform to the model were to be considered transitory, irrelevant accidents. But if for a moment he stopped gazing at the harmonious geometrical design drawn in the heaven of ideal models, a human landscape leaped to his eye where monstrosities and disasters had not vanished at all and the lines of the design seemed distorted and twisted."

-from *Mr. Palomar*, by Italo Calvino

Table of Contents

Preface

This book is a personal account of how mathematics and non-mathematics interweave. It is a collection of essays connecting my experiences as a mathematician to my other life experiences. I describe ways that my diverse life experiences have informed my approach to, understanding of, and process of doing mathematics. At the same time, I try to show how math and its various structures have helped me understand life. My perspective is that of a 26 year-old, second-year graduate student that has just passed qualifying exams.

For non-mathematicians, my goal is to tell stories of what it's like to do math – to learn, understand, create/discover, communicate, and teach math. It turns out that higher math is quite different than grade school math. The average person probably doesn't know much math past calculus, but calculus is now 300 years old, and a lot of things have happened in that time. At its higher levels, mathematics is more like an art.

For mathematicians, my goal is to make the case that process matters in mathematics – that we should actively investigate our creative processes, for the sake of becoming better mathematicians. We should examine the culture of mathematics, and the ways it has shaped and been shaped by the content of mathematics.

Part I addresses the connections between mathematics and the other dominant threads of my life. Part II contains several more minor correlations I've noticed. Part III examines the culture of mathematics. Part IV looks at the big picture and looks ahead to the future evolution of mathematics. The final chapter pulls together the various threads and patterns of my story to suggest a dramatic direction for the next mathematical revolution.

If you read this book, you'll learn some things about math, but you'll also learn about non-math. There are descriptions and discussions of different world rhythm traditions, Indian philosophy and religion, complexity theory, Buddhist meditation, and other topics.

I apologize if any of the information I present is incorrect. I'm not an expert in anything, but I'm afraid I'm too young to know better and keep my mouth shut. You'll notice that there are lots of stories; I apologize for those that seem uninteresting or too tangential.

You don't need to know any math to read this book. Occasionally I'll throw out a math term like "stable homotopy theory", but don't be scared – the meaning of any strange math terms is completely irrelevant to the discussion.

I'd like to thank EM for inspiration, RT for motivation and MJ for being so weird. The book is dedicated to two heroes of mine: Hermann Hesse and Paul Newman.

Part I: Five Vertices

Math and Music: One, Two, Three, Five, Many

People often comment on the connections between math and music. I've spent almost the same amount of time pursuing music as I have studying math, and here I'll draw some parallels from my experience.

One

North Indian rhythm theory is the most complex rhythm tradition I've been able to find in the world. The rhythms are played on a pair of goatskin drums called the tabla. I started studying tabla in college, and studied seriously for five years. My teacher in Philadelphia, Lenny Seidman, was a student of Ustad Zakir Hussain, considered by many to be the best tabla player in the world today. (I've been lucky enough to meet Zakir a few times, and even had lunch with him once.) I've now traveled to India four times, and have spent overall almost two years there. On my trips to India I always carried my tabla, and ended up playing with many musicians and studying with two other teachers.

Whereas in the West most music is in cycles of three or four, in Indian rhythm theory there are over a hundred different talas. A tala is a rhythm structure with a certain number of beats (ranging from 3.5 to 128), but with additional structures within those beats (for example, a piece of the cycle in which all sounds must be "closed", and dominant and subdominant beats).

There are about a dozen sounds you can make on the tabla, and each one has a corresponding spoken or written syllable. Most of the teaching and learning is done orally, with the general rule being: if you can say it, you can play it. Here is an example of a theme in tintal, a 16-beat cycle.

```
Dha tit Dha ge na Dha tere kita
Dha tit Dha ge Ti na  ki   na
Ta  tit Ta  ke na Ta  tere kita
Dha tit Dha ge Di na  Gi   na
Dha
```

Most compositions have a theme-variations-ending form. The variations pick out various phrases from the theme and manipulate them: repeating, augmenting, truncating, altering, rearranging, stretching, compressing, etc. When you are first learning these variations, there is no improvising, but as you progress (traditionally after four years) you are permitted to experiment with creating your own variations. This is harder than you'd think, since within this experimentation you must adhere strictly to all the additional rules that accompany the particular tala you are playing in. As with all improvisation, it is this underlying structure that gives the improvisation form and depth.

For me, the experience of improvising on a tabla theme is very similar to the experience of pondering math.

Pondering is not something that mathematicians do all the time, and I don't have the time to do it nearly as much as I'd like. I'm usually busy working on particular problems, or focusing my mind on understanding one particular topic, or trying to completely avoid thinking about math. But when you're pondering, there is no goal or destination. You are simply playing around with whatever mathematical objects are in the front of your mind. You rearrange, turn upside down and inside out, contrast, juxtapose, permute, zoom in, zoom out, and generally do whatever you feel like. Of course, you are always restricted by the massive edifice of established mathematical truths, in which no inconsistencies are allowed. I think this is a form of learning, or relearning, since after a good ponder I always feel that I better understand the ideas involved. A mathematical idea that hasn't been pondered is like an unopened present, and remains a mystery.

The same is true with a tabla composition. The theme is just a quick sketch of the musical idea, but in the variations and elaborations you really can go somewhere and find new things.

Both these activities, pondering and improvising, are best done when I'm alone in a quiet, peaceful place. They both have that precious deeply personal quality, of giving my own mind the space to be free. They account for some of my most cherished and beautiful mathematical and musical experiences.

Two

Traditionally in India the relationship between teacher and student is one of the most important relationships in life. Some say it is the most important; once I was told, "Your Guru is above God, because your Guru shows you the way to God." I've been fortunate to have excellent tabla teachers that were flexible and accommodating of my learning style.

When learning tabla, my favorite game was to try to figure out the variations before they were finished being presented. For example, Lenny would play the theme repeatedly, and once I saw what he was playing I would join in. After he was satisfied that I understood, without a pause he'd say, "OK, first variation..." and I'd stop and listen. I'd try to figure out the idea behind the variation, for example: take the first phrase in the theme, a phrase of three, and repeat then truncate it in a 3+3+2 pattern, then conclude with the second half of the theme, then repeat all that but closed for khali. My mathematical mind enjoys recognizing patterns and parsing information, and over time this ability became coupled with rhythmic intuition about what might be going on. On a good day I could figure it out and join in before Lenny was finished presenting it for the first time. The result looks and feels like mind-reading, and it was always a magical experience for both me and my teacher. One teacher in India once told me, "Wow, Luke, you eat material like a monkey."

There is an experience similar to this in classical Indian dance accompaniment. In Philadelphia and in India I got the opportunity to accompany Kathak dance in classes and performances. Kathak has a rich tradition of moves and symbolism, and there is a language similar to the tabla language for describing sequences of movements. The languages are not identical, so there is room for interpretation. At a particular point in a Kathak performance called tarana, when the tempo is quite fast, the dancer will step up to a microphone and recite a sequence. The tabla player must immediately interpret, memorize and process the sequence, and play it right back as the dancer dances. Often you can't get the whole sequence on the first pass, but you must get the main phrases and ideas, and be able to play close enough to this skeleton while maintaining the beat and all the structures of the tala. Then the dancer will recite a new sequence, or maybe this time the drummer will offer a sequence and the dancer will have to immediately dance it. This back-and-forth continues, lightning fast. Watching it, you can almost see the electricity and sparks flying. Doing it yourself feels like a transcendent mind-meld sort of communication between people.

I'm telling these stories because they are the best one-on-one musical experiences I've shared, facilitated by tabla. I've also had similar one-on-one mind-meld experiences facilitated by mathematics. Here's an example.

Last spring I did a reading course with Monty McGovern based on William Fulton's book on Young tableaux. I would meet Monty once a week to discuss the material I had read that week, to ask questions and clarify important ideas. What made these meetings so exciting is that, like the tabla teachers, Monty seemed to assume that I had an infallible ability to instantly process new information. I don't, but I like being presented with that challenge. We would talk for 20 or 30 minutes, and the whole time sparks would be flying and I'd be on the edge of my seat. It got really interesting later in the book, when Fulton goes off the deep end and uses Young tableaux to characterize objects in algebraic topology and projective algebraic geometry, which I knew nothing about. I did my best to cover the requisites from these vast and complex fields, and was able to grasp the main themes and important points of those chapters, while remaining clueless about most of the smaller steps.

Three

I spent a year in college studying Cuban bata drumming. In a bata ensemble there are three drummers, each playing a double-headed, hourglass-shaped drum. There are three different sizes. The largest is the mother drum, sitting in the middle, and the father and child drums are on either side. Bata rhythms are complex in a totally different way than tabla, which is always performed solo. They are tightly woven polyrhythmic fabrics in which songs and conversations between the six heads can be heard.

My bata partners were Sorelle, a math major who had spent time living and drumming in Ghana, and Morgan, a hip-hop and African dancer with incredible drumming skills. One of the first things we learned about bata is that it's almost impossible to practice alone, since the rhythms are so tight and the music is so much in their layering. Without the other two drums, playing is either too easy or too hard. Bata also continually juxtaposes cycles of 2 and 3, and is notorious for dragging or jumping the beat as much as possible, so most of the time the music is kept on the edge of a chaotic collapse.

The beauty in bata music is how emergent melodies ride on top of the chaos. Most rhythms have a song attached to them, and they are like rainbows – not in the air or the rain or the sun, but emergent in their cooperation. There are also various conversations between the drums, calls and responses that can be heard by a discerning listener.

I want to compare this collective experience to the way it feels to do math with a group of people. The most common instance is in a study group, when a bunch of sleep-deprived and clueless grad students get together to work through problems and cut their workload down.

At first my study group experience was dominated by frustration that I was in school on the weekend or in the evening, dread that I was so stupid I couldn't answer any of the problems, and terror at the amount of work before me. Before grad school I'd never done math in groups, and didn't know how. I think most people at the study group were thinking the same things.

Over time I came to really enjoy study groups. (It even crossed my mind that the reason they assign so much work is to force us non-geniuses to learn to work in groups.) Study groups are filled with surprises. Problems that seemed impossible to each individual at the table seem to solve themselves. Through our various small contributions and perspectives we're able to piece together solutions to the most formidable problems. Or problems I thought I understood turn out to have more depth or subtlety to them, and by working through this I understand everything better. The whole is definitely greater than the sum of the parts.

One interesting thing about study groups is that often the most banal roles become key. Tophe Anderson always liked writing on the board, and the simple act of writing down things that had already been said allowed us to collectively see the way to the solution much faster. If I had a role in study groups it was usually either to say stupid things that were mostly wrong (if I thought of them myself), or to simply reiterate and rephrase what other people were saying ("so you're saying that…"). Believe it or not, this was helpful to the group sometimes.

This is also the case with bata rhythms. The smallest drum often plays a simple and repetitive 'ee ah – ee ah – ee ah – ee ah –', but it's unbelievable how profound this piece is in the whole rhythmic fabric.

I want to try to describe how it feels during these experiences. Sure, after a study group session I feel grateful and happy, and briefly wonder at the magic that has transpired. But in the thick of it, there isn't that larger perspective. It is as though my mind is hijacked by some larger entity; the images and lines of thought that come in and out of my mind are not mine, nor do they belong to the person talking or at the board. They are shared, but not equally. While our minds seem to overlap, it's clear that we're each personalizing the discussion while we're creating and sustaining the discussion using objective symbols and dialogue. The objective and subjective dimensions of mathematics blend seamlessly. In fact, I'd even say that the distinction breaks down and I can't tell where my mind ends and the collective mind begins. This mental space transcends the self.

Exactly the same thing happened when playing bata. In the heat of the moment, deep in the rhythm, I'd look down at my hands moving, then at Sorelle and Morgan and their hands.

There seemed to be no volition; the rhythm had entered us and was playing through us. I would smile at Morgan and she would smile back, then embellish the rhythm by adding a call to the fiery beat. Without knowing what was happening, my hands would play the response, which would mesh perfectly with Sorelle's right hand's driving upbeat.

Before I found out about bata or tabla, I started drumming with a djembe in drum circles. In fact, for a full year at college I averaged 20 hours of drumming a week, and most of this was in late night drum jams. I also accompanied African dance classes, and would recruit drummers and dancers to join me out of the studio. We'd meet in the bell tower and drum and drum; people would wander by and join us, people would dance and sing; people in town would call the cops and they'd come and make us stop. So we'd go down into the woods near the college and make a huge bonfire and drum and dance until we fell asleep.

It was all very primal and freeing, and we all learned a lot from those drum circles. But the musical experience wasn't as matured and tight as with bata. There were too many people playing at once, some making only noise; people often weren't listening to each other, absorbed in their own emotions and expression; talented artists often held back out of timidity; when there was a communication between drummers or between a drummer and dancer, it was tenuous and soon overwhelmed. In short, we were young and inexperienced.

Unfortunately, math study groups also aren't always transcendent connections. Miscommunication, poor expression, noise, and timidity all get in the way on a regular basis. As we've matured as mathematicians and as we get to know each other, our study groups have gotten tighter. I look forward to maintaining these relationships and watching how they continue to develop.

Five

A few of us started a Japanese taiko group while at Swarthmore College. Taiko drumming has been used in local ceremonies in Japan for centuries; it was only in the '50s that someone thought to put it on a global stage. Since then it has become quite popular. The drums are large and beautiful, played with massive sticks called bachi. The movements are dramatic, precise, and graceful, like most Japanese arts. Playing taiko requires a level of physical discipline, and the performances are staged like dance pieces, usually with some aspects of theater. The rhythms themselves are relatively simple, compared with other world rhythm traditions. The drama of simple precision and unity, combined with large visual gestures and choreographed movements, leave a powerful impression.

What does this have to do with math? If you sacrifice subjectivity and individuality for the sake of unison, you get precision. When many mathematical minds come together over time and try to agree, the personal dimensions are downplayed and math ideas that seem independent of people start to emerge. There is power and universality in the simplicity of, for example, the laws of arithmetic. Is it possible that the apparent objectivity and universality of math can be explained as emergent behavior from a choreographed agreement among minds? This question will be explored in later essays.

Many

My other musical passion in college was Balinese gamelan. A gamelan is an orchestra of instruments – gongs and metallophones of various sizes, cymbals, drums, and often a flute or two. A gamelan is thought to encompass not just the instruments, but the musicians who play them. The melodies are based on a five-note scale. The compositions are long and complex, with many layers of rhythm cycles. The largest gongs, for example, play maybe once every two minutes. Smaller gongs play more often. The largest metallophones play slow melodies, while smaller ones play other melodies at twice or four times the speed. The fastest melodies are on the smallest instruments, and these are faster than any one person can play; these instruments are paired off, with one person playing as fast as possible on the downbeat and the other on every upbeat, for a shimmering, lightning-fast gushing of melody. The best word I've heard used to describe gamelan music is "cosmic".

Every aspect of being in a gamelan is a collective experience. As with bata, it is very difficult to play without the whole orchestra. All parts are learned by rote imitation from the teacher, and ideally everyone learns every part so rotation is possible. The compositions are incredibly rich and complex, a layering of melodies and rhythms that stretches our human capacities of attention and pattern comprehension. Listening or playing gamelan music, you can't catch all the patterns and structures, and you know that you're missing some or most of them. And yet there's an incredible intimacy with the music, since everyone knows every part, consciously or unconsciously.

I think of gamelan as taking the emergence of bata and the universality of taiko to the next level. Because of its complexity, it is more like higher abstract mathematics. And just like higher mathematics, gamelan has developed a mythology to accompany the music. It is believed that the gamelan music exists objectively and eternally. Every piece is begun and concluded with a strike of the largest gong. That first gong sound is the beginning of the piece at hand, but is also the last beat of the last time this composition was performed. Thus the beat never truly stops, and is sustained through time by being manifested from time to time with a particular gamelan. The definition of gamelan is often extended to include not just the orchestra and musicians, but also all the ancestors of all the musicians, and any other musician that has played on these instruments, and their ancestors as well.

The mythology of mathematics, carried since the time of Plato, is that mathematical ideas that we discuss and play with are Platonic ideas existing eternally and independently of the human minds that sustain them. The numbers and equations we write and speak are merely manifestations of eternal principles. Most other branches of philosophy of knowledge have abandoned Platonism, but it is alive and thriving among practicing mathematicians. While I don't intend to address the philosophy of mathematics directly in this book, there is a common undercurrent of philosophy in all these essays. My view is that mathematics is an aspect of human culture, and as such is contingent as well as universal. Over millennia, for reasons that I will discuss later, it has developed a mythology of transcendent truth, quite similar to the mythology of Balinese gamelan music.

Our gamelan group once toured with the Philadelphia Philharmonic. We performed a few shows in Philadelphia, and then went to Carnegie Hall in New York City. Dressed in traditional Balinese costumes, playing on our ornate and colorful instruments, we had the house lights left on so that we could engage the audience with our energy and smiles. When we were finished, the lights dimmed and we were replaced by the austere tuxedos and gowns of the philharmonic. I had the honor of playing the largest gong during those gamelan performances, and I'll never forget the sound of that first and last gong strike resonating in that hall.

Math and Nomadism

The Appalachian Trail is a continuous 2,174-mile hiking trail that follows the Appalachian mountain range from Georgia to Maine. Every year roughly 2,000 people start in Georgia with the intention of walking each mile up to Maine. About 200 make it all the way each year.

Swarthmore College has an optional Honors program: the last semester is spent reviewing all the material learned in your major and minor, and in May there are three four-hour written exams, created by external examiners based on your individual course of study. After 10 days there are three hour-long oral exams with those examiners, who are brought to the college. Ten days later you (hopefully) graduate, with some level of honors to your degree. It's a chance to synthesize your college studies and take some very hard tests.

I decided to study for my Honors exams – two math and one physics – while walking the Appalachian Trail (AT). I started on February 8th, 2004, on Springer Mountain in Georgia. The first morning I woke up with snow on my sleeping bag, and for the first three weeks hiked in snow. I lost feeling in my fingers and didn't regain it for about a month and a half. I'd walk 10-15 miles each day, sleeping in shelters along the trail, joined at night by a few other hardy hikers. As it warmed up, we worked up to 20-25 miles a day, minimizing our packloads to the barest essentials. (If it only rains once a week, it makes more sense to walk in the rain once a week than to carry rain gear every day, right?)

I carried math and physics books – but not all of them at once. Before leaving I chopped them up into chapters and grouped each chapter with old homework assignments. I had a "bounce box" – a large box full of all my books that I would mail to myself. Every other week I'd hitch into a town, walk to the post office and retrieve my bounce box. I'd sort through it and choose the next two weeks of material, then pack it up again, look at a map, and mail the box to a town I'd reach in two weeks. I only ever carried 5-10 pounds of papers, but this is a lot when people are cutting their toothbrushes in half to save weight.

Each night after food I'd huddle in my sleeping bag with my headlamp and pen, working through problems and reading material. During the day I'd walk and think and ponder. The problems would stretch out in space, unburdened by the need to be solved and free to mosey with me. They were like Zen koans, little puzzles to be chewed on until they solved themselves.

The other hikers got used to it all, but would tease me a bit. In fact, most of those people crazy enough to attempt a thru-hike of the AT had their own bizarre quirks. Some hiked barefoot, some with old-fashioned gear, and some were writing books. One ex-military guy would leave out candy wrappers by his head with a dagger so when the mouse snuck up at night he could kill it in cold blood. Besides the studying I also decided to stand on my head every night – one second for every mile I'd hiked since Georgia. I kept this up until Pennsylvania, when it had reached 1200 seconds or 20 minutes.

By early May I had walked up to Pennsylvania. I got off the trail and was picked up and brought to Swarthmore, in the suburbs of Philadelphia. In a few days I took my three exams. I took the ten-day break to go work on an organic farm in New Jersey where I had apprenticed a few years earlier. Then it was back to Swarthmore for the oral exams. In a few more days I was walking through the gates with a robe on and all my family there, having graduated with High Honors. The next day I was back in the woods walking towards Maine, but this time without books. I reached the end, Mt. Katahdin in Maine, on August 26th.

What gave me the idea that hiking and studying math might work together?   At that point it was a matter of faith.   I believed in integration and synthesis. I was sure that the most abstract and cerebral pursuits could be joined with the most physical and everyday activities, and their union would transcend the duality.

My favorite movie growing up was The Dark Crystal. It's the usual story of good vs. evil, with Jim Henson puppet characters. The vulture-like Skeksis use their army of huge beetles to reign in terror and guard their piece of the fragmented crystal. The peaceful brontosaurus-like urRu keep to themselves but are slowly dying off. But in the end, a Gelfling named Jen fulfills the prophecy and delivers the missing shard to return the crystal to its unbroken state, and suddenly the duality is transcended. The peaceful, good urRu and the violent, evil Skeksis are reunified in one people, the urSkeks, that are neither and both, good and evil.

Often in life I've strived to live such contradictions. Before starting my walk, I was confident that the AT would blow my mind. The old way I understood math, and the old way I related to my body and Nature, was going to be shown as the silly fragmentation it was, and I was going to move to a higher plane where mind and body were the same thing.

At least that was the idea. On the AT it never really happened like that. The experiences were always too disjoint. I'd walk and look around, not thinking but observing silently. Then at night I'd close out the rest of the world and focus on the math. If I was walking and started thinking about math, the mathematical images that filled my vision would block out the forest and trail, the statements and dialogues that would start up in my head would drown out the sound of the wind. I might alternate between these worlds, but they stayed completely disjoint in their flavors and sensations.

At the time I realized this was happening. I wasn't discouraged, for it was all just an experiment to see what would happen, what it would feel like. I didn't know what to expect, and knew that union like in The Dark Crystal was hard to bring about intentionally. But I did wonder if it was just a silly dream I had, of the mind and body uplifting each other to a new whole.

In other ways, I did feel that the experience helped me understand the mathematics better. It was clear that I was making progress, and in the end I did quite well on the exams. What was it about being outside that helped me process and study? While the experiences of studying and walking stayed disjoint on the conscious level, unconsciously they were informing each other and changes were taking place.

After the AT, I continued with a nomadic lifestyle. I worked for a bit to save money, first in construction and then for the 2004 presidential election. Two days after Bush was reelected, I headed to India to undertake a strange social work project I had concocted and labeled *The Human Kindness Experiment* – an attempt to counteract the terror and negativity that America was spreading around the world at the time. I'll describe this project in the essay *Math and Compassion*. While I was over there, wandering from city to city, changing and learning so much every day, many new strange things happened with my mind. I was out of college, and for the first time in a long while was not a student. At one point I started wondering about how my guts worked, literally – how the stomach, liver, kidneys, intestines all fit together and what they did. My curiosity grew and grew, until I couldn't help it and hunted down an anatomy book to look at. Eventually other things popped into my head, things I'd always been curious about.

India isn't the easiest place to find well-written English nonfiction books. But I found them. And Amazon will deliver anywhere.

I read about ecology, developmental psychology, network theory, Buddhism, and consciousness studies. I hiked up in the Himalayas with microbiology books. The Ganges River emerges from a glacier in the mountains, called Gomukh or "the cow's mouth", and I walked there reading about Deep Ecology. I left India for New Zealand and hitchhiked, trekked, and farmed around New Zealand reading several books on complexity theory. I got a job on a sailboat and sailed from New Zealand to Japan, stopping at New Caledonia and the Solomon Islands, and on the two-month journey read a lot – mostly world history. After a couple weeks in Japan, I took a ferry to Shanghai and traveled overland through southern China for a month, up to Chengdu and then to Lhasa in Tibet. Oh, Tibet. I fell in love with this magical land – its mysteries and its tragedies. During this time I read a few history books about China and Tibet; one of them was published in Shanghai and full of propaganda, and one of them I had smuggled in from Japan. I hitched and trekked through the villages of Tibet, including a trip up to the Everest Base Camp and beyond it to 6000m. I hitched over the Himalayas down to Nepal.

After a trek in Nepal in spite of the monsoon (I walked from teahouse to teahouse carrying a daypack with just a Kurt Vonnegut book and an umbrella), a few buses took me through Nepal back to Rishikesh, my favorite town in India. Rishikesh is a sacred pilgrimage town where the Ganges comes out of the Himalayas and meets the plains. I stayed in Rishikesh for six months, doing more social work and more treks, reading mostly Indian philosophy and Indian music theory texts.

Through all of this I carried only a small pack, but I always had two or three books. I'd pick up new ones and leave old ones behind. Most of them were nonfiction; every nonfiction book was not just read, but read carefully, and even outlined! All the important points and themes were summarized, diagrams were drawn, and reactions and thoughts were documented, in a growing collection of notebooks.

I also took the time to learn some of the basics of the various languages: Japanese, Mandarin, Tibetan, and Nepali, which I promptly forgot as soon as I left. I'd already learned Hindi on my first India trip but I taught myself more, to the point of very rough fluency.

From India I went back to New Zealand, for more hitchhiking, trekking, and working on farms for room and board. I ended up with a job picking apples for a few months while staying on a socialist commune, reading about neurophenomenology and transpersonal psychology. I left New Zealand on the first day of winter, June 21st, 2006, and after a night out in Hong Kong arrived in New York on the first day of summer, June 21st, 2006.

Almost two and a half years passed from the time I started walking the AT to the time I returned to New York. Innumerable experiences from that time have changed me in ways I never imagined possible. Humbled by the vastness of our earth and our human condition within it, every day I am grateful. Condensing so many travels and studies into a laundry list of a few pages only accentuates that vastness and depth. In the end, the richness of existence is everywhere and in everything, and I'm not sure it's necessary to travel so far to find it. Still, I try never to forget how privileged I am to have traveled and experienced the things I have. Being a tall, white American male from an upper-middle class family has allowed me to safely and (relatively) easily go almost anywhere I want. All I can do is be grateful and try to take advantage of my opportunities as much as I can, while helping to give others the same.

One important lesson from that time, for example, is that curiosity is a magic key that gives us the power to unlock the world around and inside us. Curiosity is a precious gift to be nurtured and cherished, and cultivating wonder should be a top priority in education.

Curiosity, creativity, spontaneity, flexibility, perspective – these are all things that I learned while traveling. These qualities are extremely useful in all of life, including mathematics. Yes, these travels made me a better mathematician, I have no doubt. For example, the diversity of travel fosters creativity and wonder. How can you be curious if everything is static, predetermined, and safe? The space, and the variety of experiences, offer perspective on problems and suggest new approaches. Every mathematician recognizes the value in conversing with other mathematicians. Traveling is a conversation with the world.

To be a nomad is to approach life such that you are always traveling. Everywhere and nowhere is home; you are what you carry with you, and this is not much. You are as free as you can be in each moment.

There are some historical precedents in the case for mathematics and nomadism. The famous Hungarian mathematician Paul Erdős was always on the road. He carried two suitcases, each half full. Mathematicians around the world gladly welcomed him into their homes, because in exchange for food and lodging he would offer some wise new perspective on whatever problems were presented to him. His intuitions were flawless, and he was able to inspire more mathematics than perhaps any other single mathematician that century. Erdős is quoted saying: "I have no home, the world is my home", and "another roof, another proof".

One of the most influential mathematicians and philosophers ever, Rene Descartes, was also quite a wanderer. In his *Discourse on the Method,* he writes:

> I entirely abandoned the study of letters. Resolving to seek no knowledge other than that of which could be found in myself or else in the great book of the world, I spent the rest of my youth traveling, visiting courts and armies, mixing with people of diverse temperaments and ranks, gathering various experiences, testing myself in the situations which fortune offered me, and at all times reflecting upon whatever came my way so as to derive some profit from it.

Most of Descartes' major work was produced during the 20 years from 1628 to 1648, which he spent wandering from city to city in the Netherlands, never staying in one place for more than two years. He claimed his best thinking was done while lying in bed late in the mornings. Famously, he died of pneumonia in 1650 after agreeing to stay in Stockholm and teach the Queen of Sweden, who insisted he wake up before sunrise for their daily lessons.

Let's return to my AT hike and that experiment in fusing math and nomadism. It's clear that traveling and diverse life experiences are valuable to a mathematician's life, but is it only a second-order effect? Is traveling helpful to math like sleeping is helpful to math, or is there some more direct way that the two experiences can be unified?

The first step towards getting a math PhD is passing qualifying exams. You have two years to pass three exams, which are given at the end of summer each year. After my first intense year of classes, I found myself with a summer to study for two exams. I was very excited. This was a chance to go back and slowly review and reprocess all that I had covered earlier that year at a breakneck pace. I was now free to take the time and build my own personal understanding of the material, to make it mine. I was also free to structure my summer studies however I wished.

For nine months I had been forced to sit in the same uncomfortable, right-handed chairdesks in the same room every Monday, Wednesday, and Friday for the same three 50-minute periods, taught by the same professors. All that time I knew: this is not how I learn best, this is not my learning style. But I was forced to do it their way. Now I was free. They had made mathematics flat and empty. Now math once again projected into that other dimension, the human dimension.

So I did my math studying in the grass, or sitting by the lake. I organized my days to allow for the long silent hours alone without distraction, letting the math juices flow whenever and wherever it was that they seemed to flow. This was complemented by biking, hiking, rock climbing, cooking food, dancing – all the things that refresh and sustain my simple life. The breakthroughs started coming and didn't stop; as I reconstructed my courses from the ground up they became mine, and my understanding widened and deepened.

For the second half of the summer I went to Sweden and Norway. Liz McTernan, a hometown friend, had an art piece in an art festival in Sweden. I had helped her out a little on the project[1], so we went together and used it as a chance to

---

[1] Liz's art piece started with a tree being felled in Eastport, Maine, the Eastern-most point of the USA. My contribution was to calculate exactly how much time it would take the sound of the tree falling to reach Trollhätten, Sweden, and precisely what direction it would come from. At the correct time, we gathered people in Trollhätten to face Eastport together and listen for the sound. The morning of the performance, I had to adjust my calculation by about a minute,

wander a little. Of course, I was excited to bring my math with me.

The art festival lasted for two weeks in the small town of Trollhätten, Sweden, which has a river running through it and is surrounded by a forest in which trolls used to live. One day I went for a walk along the river, through the forest. My mind was clear and the air was fresh. I had been studying math earlier that day, and decided to ponder a little. At the instant of that mental volition, something broke inside like a veil lifting, and for the first time ever I experienced the unification of mathematics and nature. The math thoughts filled my mind's eye, but rather than overwhelm and block out my visual perceptions of the forest, they fused together. I heard the math narratives and discussions in my mind's ear, and they fused with the sound of the river and trees. I can't describe it any more clearly. There was no effort involved; the math structures seemed to expand on their own to meet the smells, shapes, and movements around me.

I quickly realized that this was different than anything else I'd experienced before, and when I realized what was going on I started crying. It had been three years since my original AT dream of unifying mind and body using mathematics and nature. I had forgotten it; the dream had left my conscious mind, and subsided into my subconscious. It was there as I read and outlined all those other books while traveling. It was there during the long difficult first year of empty grad school mathematics. Now it had been realized, without me even trying. The beauty and depth of the moment filled me with a profound gratitude. I stayed in this state until my walk ended.

---

because the average temperature over the North Atlantic was one degree higher than I had first anticipated, and as a result the sound would be traveling faster.

The rest of the trip was amazing. We traveled and explored Scandinavia, and I studied. I was often able to revisit this fusion of math and nature, mind and body, and in many different contexts. Everyday experience informed the math experience directly. I knew I could never go back to that flat mathematics. Math is not independent of humans! Or even of what I ate for breakfast this morning! It exists in the minds and practices of humans, and is shaped by and shapes them.

That summer I was studying for two exams: one on abstract algebra and one on topology and smooth manifold theory. I want to describe to you three contexts in which this fusion of math and non-math resurfaced.

My study in Trollhätten was mostly aimed at rereading and outlining the texts, summarizing the material and organizing it based on the central ideas. Liz and I were camping with all the other artists. Since it was summer, it was only dark for two hours of the night. I would do my studying in restaurants and bars in the town, or in the campground kitchen. I was able to get into and sustain a focused state, since the ambient language wasn't English, there were no written English words, and I knew very few people. The studying time was complemented with bizarre happenings with the artists at the festival – playing with slugs, riding bicycles into the river, crashing the local Jazzercise classes, etc. Many of the artists were using the time to create, and many of them were conceptual artists with unconventional styles. I had many excellent conversations with these artists about math and art. Although I never came close to describing the subtleties of the problems I was working on, these conversations carried that thick immediacy of being math and not math at the same time. The ideas thrown around informed my mathematics directly it seemed, most often by restructuring the ways I was visualizing the concepts. This was not only a unique type of exchange, but was extremely productive for me.

Liz and I left Trollhätten and headed west to Norway. In Norway, as in all of Scandinavia, they have an ancient tradition, now written into law, of "freedom to roam". Throughout the country, anyone can camp anywhere, on public or private land, provided that they are away from roads, out of sight, and that they clean up after themselves. So each night we camped for free in the woods. We hitched to the fjords, where we did a lot of hiking. During this period we moved around all day, every day, and it rained a lot. At that stage of my understanding of manifold theory, the main ideas were still burdened and obscured by lots of technical details. It was hard for me to make progress on problems without a pen and paper and time to fiddle around. So I couldn't study manifolds very effectively. On the other hand, the algebra problems were like koans or puzzles of just the right size. They were perfect for standing by the side of the road with our thumbs out. I would keep some papers with problems on them in my pocket. The stunning natural setting we were in, coupled with the random spontaneity of hitchhiking, made for just the right amount of stability and chaos so that the algebra structures could be sustained and dissolved as necessary to effectively find solutions. Likewise, when we went on hikes I would slip into a state similar to my initial Trollhätten breakthrough, and the problems would unravel themselves and recombine in solutions as the clouds in the fjords would move and transform with the topography.

From the fjords we went east onto the high plateau. We took one three-day backpacking trip during which, mathematically, I faltered. The setting had changed; we were now crossing vast alpine meadows and tundra, moving among icy peaks and their glaciers. I had been dislodged from the math groove of the fjords. Those techniques didn't work any more. I knew I had to change my approach, but didn't know what to study – or how to study – next.

Enduring my anxiety, I decided to wait and see. I stopped doing math for a few days. We returned to a town and got a week of supplies, then headed up a valley to set up a base camp from which to do day hikes. Our hikes would take us up side valleys to the base of the glaciers, where we would stop and sit silently for a long time. I wouldn't think about math at all, or really think about anything, during those trips. We climbed a nearby peak, the highest point in Norway. That night I decided to sit down and try some old exam problems again. The exams have eight problems, and getting four complete solutions is considered a pass. You're given four hours to do this. I turned to an old manifolds exam and started it. Suddenly, I had six complete solutions done, and an hour and a half had gone by! It felt like brilliance had struck me. I turned to some old homework problems from the text. I effortlessly completed all the old homework problems from one chapter, then turned to the next chapter and did the same, then again with a third chapter. I can't remember ever doing math so quickly and effortlessly. So this was how I was to experience math in that silent, vast valley of glaciers – like lightning striking. The lightning continued until we hitched out of the valley and headed back to Oslo.

What's the point of these stories? I'm attempting to express and demonstrate how my math experiences are directly and immediately correlated with my state of being, my environment, my interactions, and my life in general. What does this mean for mathematics? We must work to understand *how* we do math, so we can understand how we do our best math, so we can be better mathematicians. Living with all those artists in Trollhätten, I was really impressed again by how much math and art are the same. Mathematicians are artists.   Math is an art.   What is different is that artists recognize how essential self-reflection and self-awareness are to the quality of their work. An artist strives to understand his or her process. This is not just so that higher quality art can be produced, but because this human dimension is a beautiful dimension. Without self-awareness in the creative process – whether artistic or mathematical – creations are flat and empty as shadows.

I hope that some day the mathematical community will value diversity and diverse life experiences. In my particular institution this is not the case, and my sense is that other institutions are similar. For example, throughout that first year, I knew that I was underperforming because my learning style was not that of the early-morning, short and frequent lecture. There were no attempts at accommodating any other ways of learning.   Also, when I decided to leave and study while traveling in Sweden and Norway, I was met mostly with disapproving condescension and skepticism. In the essay *Math and the Math Culture*, I'll discuss this lack of diversity and its possible origins.

It was during that summer in Scandinavia that I started calling myself a mathematician. When I returned to the grad school setting, my mathematical experience was once again restrained and flattened.   However, my understanding of what mathematics is and can be had expanded, and wouldn't go back.

Math, Hitchhiking, and Couchsurfing:
Malfunctioning in Society

In the summer of 2008 I taught my own class, Math 111: Business Math. It was a requirement for those entering the business program, and a lot of psychology majors took it for their math requirement. During the first week I sent out an email that included the following passage:

> "The most important thing you might learn in this class is not the particular mathematical content we're discussing. Many abstract aspects of "doing math" could make you a better businessperson, psychologist, or economist, or whatever. Mathematicians must present and organize information in a rigorous, logical, and concise way. As we solve problems we use our creativity and intuition. Doing math also requires developing communication skills, in order to go from the internal and subjective to the external and objective. So even if math isn't your thing, I hope you can get something useful out of this class."

This is what I often used to tell people about the benefits of doing math. Even if you don't use calculus in your day to day life, in calculus class you've hopefully learned things that can be applied to any situation: problem-solving, analytical thinking, how to work in groups, communication skills, etc.

I've now decided that this is a farce. Math doesn't prepare you for functioning in society. In my case, what have been the effects of doing all this math that I do?

I barely keep in contact with my family and friends outside of Seattle. I'm notoriously late for things; if the math juices are flowing then I can't help but let them, and have been known to arrive hours late or not at all to scheduled events. When there is tough math percolating in the back of my mind, I'm hardly much fun; I sit quietly and stare or wander aimlessly. It's then hard for me to engage with people, because I can't take in much new information. I become boring and serious. When all you do is math that few can understand or care to hear, then you don't have much to talk about.

The extreme focus involved in mathematical thought is also detrimental to functioning in society. Being in the world requires continual multi-tasking and parallel processing of information. When I've been doing lots of math, my concentration becomes too acute to handle more than one thing. For example, I can't drive a car and talk at the same time. Driving a car at night gives me small panic attacks. Countless cooking attempts have resulted in disaster when I've tried to hold a conversation and cook simultaneously.

For some reason, I'm usually not interested in applying my glorious problem-solving skills to bring reason, clarity, and focus to others' life puzzles. If I do try, I'm usually not very successful. Why not? People don't need carefully reasoned solutions. Life is not rational and linear enough for my skills to be of much use. What people need is empathy. They need a quiet, patient, attentive listener with an open mind and heart who can listen to what they say and help them discover their own solutions.

Math is not good for this in the slightest. After doing math, my mind is not quiet, not usually attentive, and not even necessarily very open.

What's more, math induces a lack of empathy. The more math I do, the more disconnected I feel from emotions. For the most part, all that graduate school math has made me feel are dread, anxiety, and a shallow pleasure due to the release of some intellectual tension. The moments of transcendent beauty, wonder, and gratitude are quite rare in a balls-to-the-wall graduate program.

Math has disconnected me from my own emotions. When I first started my grad program, the transition back to an academic setting, after just moving to a new city, was quite stressful. After the first week, I broke down and cried. The second week continued the stress, and at the end of the second week I also had a breakdown. After the third week, another crying fit hit me. Almost without fail, at the end of every week I would collapse into tears. I blame the stress and pressure, but also the extreme disconnect from my emotions. During the 60 hours a week devoted to mathematics, emotions were overwhelmed by reason and suppressed. The result was a weekly outburst of a week's worth of emotions, all at once. As the year progressed, my breakdowns would come earlier in the week. By the end of that first year, I couldn't make it past Wednesday morning without crying.

Spending so much time in the worlds of math keeps me out of touch with myself and my own emotions, making it harder for me to open my heart to others.

Math has also disconnected me from others' emotions. Math reasoning is frail and intricate compared to the robustness of real life. Thus the math community maintains the ideal mathematician is someone removed from the vicissitudes of everyday life. We're supposed to live a monastic life with few distractions, so our minds are calm and clear. By avoiding the struggles that my family and friends are working through, I'm losing my ability to understand their feelings and actions.

Eccentricity is almost expected from mathematicians. We are expected to lose track of social norms and the rules everyone obeys, and of why it is we're supposed to obey them.

Considering how indescribably out-there most mathematics is, insanity is almost an occupational hazard. In some contexts, this eccentricity can actually be a wonderful thing. It is very freeing to be surrounded by oddballs, with no good sense of who is sane and who isn't.

More often than not, I believe that people are their own oppressors. We keep ourselves chained down, out of fear or lack of imagination.

I remember, in Tibet, taking three days to walk up to Everest Base Camp, at 5,400 meters above sea level. There is a monastery, and a small encampment of tents. From there, nothing separates you from Mt. Everest but a long glacial basin leading up to the confluence of several glaciers in the cirque below the Everest massif. At the southern part of the encampment, there is a large sign declaring that anyone passing beyond this point without a certified expedition will be arrested and charged US$200. So I stocked up on supplies – which here meant one packet of Ramen noodles for each meal – and when no one was looking, headed south past the sign. I was alone and wearing sneakers. That night I made it as far as an area I think is called Advanced Base Camp III, at about 6,000 meters or 20,000 ft. I camped on the glacier and cooked my Ramen noodles, surrounded by incredible 40-foot towers of pure blue ice shooting straight up, called seracs. The altitude sickness made me too weak to go any further, and besides, from there I would've had to cross a scary crevassed glacier on the other side of which the mountain seemed to shoot straight up to the summit. On the way back I took a different way, and was confronted with an impassable glacial creek. I camped again and was able to ford it when the flow was low in the morning.

There was not an instant during this trip that I felt like I was operating outside my comfort zone; I was careful and aware at each step. If I went back there with any reasonably healthy person, I'm sure we could reach the same point and get to experience the same incredible landscape. We don't know what we can and can't do until we try. More often than not the limitations we've put on ourselves are insubstantial mental chains.

So while the engrossing worlds of mathematics tends to disconnect its practitioners from the social norms and societal concerns surrounding them, in some contexts this eccentricity is important. Have you ever gone dancing with a mathematician? In my experience, once they've overcome their timidity, mathematicians can dance with such inspired, creative zeal as to infest the dance floor with ecstatic wonder.

However, eccentricity quickly becomes a problem and a burden on society when communication skills break down. This is also a common possible consequence of doing mathematics. It's clear to me that when I'm in my math mode, I struggle to express myself clearly in everyday life, to the detriment of those around me. I'm not really sure why this happens; it feels like my mind gets ahead of my words, and it's very frustrating. I've been able to watch my descent into inarticulateness several times, thanks to a few activities that counteract the process. Two of my favorite things to do that promote communication skills are hitchhiking and couchsurfing.

I started hitchhiking while on the Appalachian Trail, in order to get into towns and resupply. Later, I hitched extensively around the beautiful islands of New Zealand; on my second trip there I counted 111 hitches over five months. I've also hitched in India, Tibet, Sweden, Norway, Jordan, Israel, and Montenegro, as well as to and from the mountains and ocean here in the Pacific Northwest. At this point, I estimate that I've hitched at least 250 times, and I have 250 excellent stories to tell from it. If you pressed me I could probably remember almost all of them – where it was, who picked me up, what we talked about.

There was the Israeli army jeep full of young men, blasting Pink Floyd as they escorted a tank to the Gaza Strip. There was the Norwegian whose job was to travel around the world training people to operate the machines that make cardboard. There was the ancient pickup with the Tibetan village doctor, stopping briefly to distribute pills and prayers in each town along the one (bumpy, unpaved) road in southern Tibet.

I love it. No, I've never had a bad hitchhiking experience, although a few times in New Zealand I got out early because the driver was not sober.

You never know who is going to pick you up when you're hitchhiking. When someone does, you don't know anything about this person except that they have some willingness to let you into their car and life for a little while. Each hitch is a wonderful coming-together of strangers, and an excellent test of communication skills.

When someone pulls over to pick me up, my first concern is to put this person at ease, and then to express gratitude. There is a slight danger in picking up hitchhikers, and I must in one way or another communicate that I mean no harm and I am very grateful for the ride, however far it'll take me. Once this is established, the rest of the ride we are free to find some unique form of relating to each other. Some drivers don't want to talk, some will gush to you their whole life story, and some want to be entertained with travel stories. There aren't any rules as to what must be communicated, except for those first two things: peace and thankfulness. Some rides last for five minutes (a ride from the center of a town to the outskirts can be one of the most helpful), others can last for hours and hours. Once I got picked up and ended up spending four days traveling with the driver. Once I got picked up and invited to the woman's home, where I stayed for a week house-sitting and using her car while she went on vacation.

Over time I've gotten better at hitching, and hitching has made me better at communicating generally. Isn't this the essence of a successful communication: first express peace and gratitude, then share something unique? After a trip during which I've been hitching, I notice that I'm more able to express myself and communicate. Conversely, if I've been doing math and take a trip involving hitchhiking, during the rides I can tell that I'm not communicating as well as I could.

Couchsurfing is a similar sort of coming-together of strangers. Each interaction is a unique one. Numerous times around the world I've made a friend and been invited into their home, but here by couchsurfing I mean specifically using the networking website couchsurfing.org to find people willing to offer travelers a place to sleep. I've couchsurfed in Stockholm, Oslo, Bergen, Jerusalem, Istanbul, Pristina in Kosovo, Kotor in Montenegro, Dubrovnik, Sarajevo, Ljubliana, and Vienna. The shortest stay was one night, the longest was 10 days. It's an excellent way to meet a willing local. I've also hosted a few people in Seattle. At this point, there are over 700,000 couchsurfing profiles in almost 50,000 cities of the world. Together they make up a community of willing, friendly people that believe in the importance of generosity and sharing. With couchsurfing, there's never a reason to pay for lodging or be without a friend in a city ever again.

Through hitchhiking and couchsurfing, I've had the opportunity to meet so many unique people, each shaped by their unique life experiences. Each is wise in his or her own way, and has something to teach me. The mathematical lifestyle also brews unique individuals with unique perspectives, and I think mathematicians have lessons to share with the world – for example, about imagination and freedom of thought. But there is a thin line between the wise fool and the crazy fool. A mathematical life, in my experience, can also lead one towards malfunctioning in society. As I steep in eccentricity, what determines whether I become a burden or a potential boon to society seems to be my ability and desire to communicate the lessons I've learned.

Math and Compassion

I've talked about the ways my passion for mathematics has synthesized with my passion for music. I've talked about the gradual development of a unification of sorts between the mathematics of my life and the nomadism of my life. There is a fourth major gesture in the movement of my life up to now, and that is an exploration of compassion. However, reconciling mathematics with compassion has proved more of a challenge.

Growing up, and all through college, I had a cold and unemotional character. I often wondered if I was a sociopath, because while people around me seemed to be feeling the vicissitudes of life I mostly felt nothing. I was intellectual and contemplative, interested in understanding my own mind and the nature of reality, not the unpredictable and seemingly arbitrary trivialities of the heart. I had a sense, though, that I was at some point going to have to engage society and the questions of civilization, and I had a sense that finding a place in society had something to do with understanding compassion.

Towards the end of my time in college, I had developed a long list of dreams and projects to undertake – traveling adventures, jobs to explore, things to try, motorcycles to ride. My life seemed to stretch out before me as one endless dessert table of delicious experiences to be had. But the mystery of this thing called compassion kept calling out to me from far below my ambition and selfishness. I made a decision: before taking on all my wonderful selfish adventures, I would suspend those dreams for a year. I would devote that year to these hard questions I barely knew how to ask: What is compassion? What helps? How can I help? What does it matter?

Well, I couldn't quite suspend all my other dreams, so gradually I concocted a wonderful plan to pursue my questions of compassion while also doing the things I enjoyed. I called it the Human Kindness Experiment.

I went to all my friends and family and proposed to them this: give me some small gift of goodwill, in whatever form you can imagine, and I will travel to India and deliver it to an appropriate person at an appropriate time. Look into yourself and think of something you would like to give, some message you would like to send to a human being on the other side of the world. Acting as a sort of goodwill ambassador I will befriend an Indian and give this gift on your behalf, making sure to tell the story of its giver, and making sure that it is given in a context such that it is received as a pure offering of kindness. Even before I had printed and distributed my project's proposal, I was confronted with those tough questions about compassion, and began to slowly discover tentative answers and explore new directions. How can I judge the impact of the gifts, or should I even try? What about accountability? How do I avoid establishing dependence and confirming power hierarchies? I created a website, www.hke.blogspot.com, where I posted stories of my exchanges, with pictures and any return messages or gifts, as well as commentary on my trip and the lessons I was learning about helping and making a difference. The website turned out to be one of the most beautiful things I've ever created, and I still occasionally visit it for inspiration. I suggest you look at it, for a better sense of the project and to read about all the changes and beautiful connections that emerged.

As one example, a tabla friend in Philadelphia taught me a short composition and gave me these instructions: share this with some Indian tabla player, and learn one in return to bring back and teach to him. So I had a wonderful mission: wander around India looking to befriend tabla players, and establish a friendship so that I could sit down and exchange compositions.

What does that have to do with compassion? I sustained a state of openness, willingness, and longing to connect. It was not for me, but for the sake of establishing a positive connection between strangers, however tenuous and short-lived – to bring more happiness, unity, and peace into the world.

The Human Kindness Experiment changed me forever. I concluded that the reason we exist is to help each other and uplift each other – in other words, to love each other. Nothing was worth doing unless it could be done with love and compassion.

It was quite a shock when I stopped traveling and settled in Seattle to start mathematics graduate school. I had many and strong motivations for starting a PhD program in math, but from the start I saw almost no connection with my new devotion to compassion. Math and compassion do not naturally go well together. As I discussed in the last essay, there is much about mathematics that interferes with or even negates a compassionate lifestyle. This set up a contradiction within me and my life that was so stark, close friends and family even commented on the conflict. I have not resolved this paradox yet, although I've made some progress.

I've tried to counteract the uncompassionate effects of mathematics. One extremely helpful tool is meditation. By meditating I can keep in touch with my deepest, subtlest self. I give a space and time for the quiet dances of my emotions and thoughts to play out and establish dynamic equilibriums. This frees me to open my heart to others. I can be aware of my own human workings and thus empathize with those around me. I can move through the world with willingness, patient clarity, and serene strength, ready to embrace and share.

While math always calls to me to leave this world for that rarefied universe of concepts, I've made an intentional effort to stay in this world. I try to keep up to date with international news, and try to stay involved in the lives of my friends and family as they change and grow.

Volunteering is also a helpful tool. I've volunteered for several community events and done some hiking trail maintenance. My most powerful volunteer experience in Seattle has been at the ROOTS Young Adult Homeless Shelter, which provides meals and a place to sleep for 18 – 25 year olds. In the summer of 2008 I volunteered over 120 hours there – enough to really get to know some of the guests. Whatever time and effort I put into helping the shelter operate, I got back so much more from the guests. These young people are so genuine, so present, and so hardy in the face of some serious challenges. I learned a lot from them, and was also able to share some lessons from my life. Helping at the shelter three times a week definitely helped keep my math studies in a wider context.

Am I bound to always struggle with the conflicting directions of mathematics and compassion, or is some integration possible? Are there dimensions of mathematics that might support or even encourage a compassionate lifestyle? I think that compassion is a human experience with many facets, and I do think that there are ways that doing math can make you a better, kinder person.

In the most immediate way, after doing math for several hours and then stopping, I sometimes feel very calm and peaceful. For a little while, before the speed of the world around me again fills my mind with chatter, I'm aware that I can exude a cheerful serenity.

Mathematicians are some of the most humble and selfless people I know. While there are definitely a few driven by ambition and ego, they stick out clearly among us. Most of us realize quickly that we aren't the geniuses we thought we were. In math, there is always someone much smarter than you, or at least more studied. Because every subfield has a language of its own, even if you are an expert in one area, you have no option but to defer to another's knowledge when you leave that area. There's no faking it in math conversation. Furthermore, in learning math we follow and retrace all the "Eureka" moments of past masters; this is humbling but also shows that as we do our own math we are merely standing on the shoulders of giants. I can't be too proud of a result I might establish, no matter how monumental, because it is just a small step beyond what others have done.

For the same reason, mathematicians have deep respect for collaboration and understand how crucial cooperation is to our art. Even those that lock themselves away in the attic for years and return with results must submit those results to be judged and approved or disapproved by their peers. Centuries ago mathematicians worked more in isolation, but nowadays with our tightly-interwoven global society, collaboration is the norm. This is of course the way it is with much art today. Art, in its creation, has the power to bring people together.

There's a good story about this sort of mathematical cooperation. For the last hundred years Russia has been a mathematical powerhouse. During the cold war, however, communication with the Russian mathematicians was difficult; most of those mathematicians worked in isolation without support behind the Iron Curtain. In the '80s, one American mathematician, Robert MacPherson, had been traveling into Russia to collaborate. In 1991, when the Soviet Union collapsed, the Americans saw an opportunity to finally extend a hand to the Russian mathematicians. Macpherson organized the raising of $100,000 through the American Mathematical Society – but it wasn't clear how to deliver this money.

Corruption was rampant at the time, and due to inflation people were being killed for a mere 50 U.S. dollars. In the end, Macpherson went over in person and smuggled much of the money in. He organized a meeting in a Moscow public school, and Russian mathematicians – many soon to become recognized around the world for their brilliance – convened to collect a few hundred dollars each.

Another, somewhat peculiar, way in which mathematics helps society is by forcing its practitioners to swallow the existential pill. Higher math is unflinchingly arbitrary and absurd. Every day I think to myself: How and why am I so passionate about this totally pointless activity? Everyone I know well, mathematician or non-mathematician, has at some point asked this question. What's the point? Why should I get so involved in the arbitrary games of life, when in the end they don't matter at all, and we all die alone and empty? I think mathematics, especially the more abstract and advanced it gets, is an extreme example of this absurdity in our lives. If you're going to stick with math and make it a central part of your life, you must somehow begin to confront this existential paradox. This aspect of mathematics *does* help us mathematicians to empathize with the people in the world around us. It is an emotional and philosophical challenge that everyone must face at some point, and doing mathematics forces you to face it directly at a young age.

The most profound way in which math itself might help us develop compassion, I think, is in what mathematics has to say about the relativism of truth. A mathematician is always working within a framework of axioms, taking some things to be true and building on those. Each area of math is perfectly consistent within itself, but between different systems there are often apparent contradictions. As a simple example, in the most common number system, we have $1 + 1 = 2$. But there is a perfectly valid and quite useful system in which $1 + 1 = 1$; this is the Boolean logic that is the basis of all circuitry in all electronic equipment, for example. In another system that shows up often in mathematics, we have $1 + 1 = 0$.

No one can say which system is right, because none of them is right. They are all different truths, perfectly valid and consistent within themselves. Some may seem more useful or more interesting, but this is a subjective judgment. If I like traditional arithmetic in which $1 + 1 = 2$, then meet a mathematician that has spent most of her mathematical life working within the system in which $1 + 1 = 0$, I cannot judge her as better or worse, or more or less correct than me. The only way to understand her, the work she has done, and the language she speaks, is to suspend my truth and step into her shoes. If I do this in earnest, a whole new universe of math will open up to me. Again, it may be impossible to reconcile these two universes. The best I can hope for is to realize a transcendent truth: that we are both humans and both mathematicians, working and struggling with our own challenges. Even though I probably can't understand the depth of her world as she does, I can respect it and call her my sister.

This is tolerance, leading to acceptance, and possibly even to a unity larger than our selves.

I think this is a universal lesson that needs to be taught today, and will need to be taught forever. I've always thought that if there were one Absolute Truth, it would be this: There are No Absolute Truths.

On September 11th, 2001 I was in Jaipur, in North India, in a study abroad program. I was playing tabla in our neighbor's flat for a birthday party, when someone rushed in and turned on the television. Of course we were shocked, but it was late and we soon went to sleep. The next morning, everything was uncertain; the organizers of the program told us to be very vigilant, and to absolutely *not* go to the Muslim area of the city.   So that night I snuck out and walked into the Muslim area of Jaipur, all alone, as yet unable to speak much Hindi. I didn't end up getting killed or abused in any way. Rather, I was taken in by some shop owners and given some delicious fresh-roasted chicken.   (Fresh-roasted chicken in India can be quite fresh. If you're lucky, they'll kill, skin, and roast the chicken before your eyes.)   Someone who spoke fluent English was fetched. It turns out that these Muslims in Jaipur had nothing but compassion and concern for my own safety. They recognized that on a global scale something was going to happen between our cultures, and expressed continually that neither side should make generalizations about the other. Rather, they were concerned and curious: Was my family in New York City? Had I contacted them to make sure they were safe? What did I think was going to happen next? It was time to start trying to understand each other.

Relativism is something that hits you in the face when you travel throughout the world. India, which I've visited four times, is home to rich subcultures, each with its own system of truths. Each regional way of life is unique – and is different from New Zealand, or New Caledonia, or the Solomon Islands, or Japan, or Nepal.

I spent two months in China and Tibet, and I fell completely in love with Tibet. Having heard the usual story of Tibet's oppression at the hand of the Chinese, I wanted to side with the Tibetans and was tempted to demonize the Chinese. By reading history books and talking with Chinese people I realized it wasn't that simple. First of all, you can't blame the Chinese citizens for being complacent: the level of government propaganda and control of information is scary, and besides, you could make the case that in America it isn't much different. There are reasonable historical arguments for why the Chinese might consider Tibet to be part of China. The Chinese have tried to modernize Tibet, as you might if you visited part of your country and found no paved roads, no reliable health system, poor hygiene and education, and high unemployment. When you saw the elite ruling class of the monks, it's very reasonable that with a little communist zeal you would try to deconstruct what you saw as oppression and inequality. In the end, the only argument I can make against the Chinese is that they did this modernization a little too fast, and in a way that wasn't as culturally sensitive as it should've been. This is a weak and subjective criticism, far from the harsh one-sided views I first held against the Chinese.

In early 2008 I took another trip that opened my eyes and my mind to the relativism of truth. From Tanzania I flew to Dubai and then Egypt. From Egypt I traveled overland to Jordan, Israel, back to Jordan, then Syria, Lebanon, back to Syria, then to Turkey. I continued from Turkey through Bulgaria, Macedonia, the new country of Kosovo, Montenegro, Croatia, Bosnia and Herzegovina, Serbia, back to Bosnia, then Slovenia, and finally to Austria. In some countries I stayed only a day, in others for three weeks. The whole trip was four months.

These are obviously some tense and conflict-ridden parts of the world. At no point was I able to take sides and claim I had even an inkling of a solution to anyone's problems. Rather, I focused on listening to each story as it was presented, trying to understand the rationality and consistency in everyone's perspective. Everyone strives to have a consistent worldview, I think; it's just that we start with different assumptions and grounding experiences. What I found was that everyone was acting reasonably, meaning that I could put myself in their shoes (at least a little bit) and say: Yes, I might do the same thing.

For example, in Jordan I went to the desert at Wadi Rum, and spent a few days wandering across the sand and scrambling around the rock formations. At one point I stopped into a Bedouin tent; I had brought my lunch and was looking for a place out of the sun and wind to eat. The man there offered me tea and gave me a place to sit on elaborate carpets with a low wooden table. He asked where I was from, and I told him the USA. He looked me in the eye and immediately said, "Oh, we hate Americans". I laughed nervously and asked why. He explained that, "First America destroyed Palestine. Then they destroyed Iraq. What will they destroy next? Maybe they will destroy Jordan." I didn't know what to say, and ate in silence. Gradually, I asked him questions about himself. He was the first in his family to go to school; he was studying computers, and the internet allowed him to see directly what was going on in Iraq. He seemed to warm to me a little. In the end, I meant it when I said to him, "You know, I think that if I was a Bedouin raised in Jordan, I would hate America too." He nodded and conceded that it wasn't the American people that he hated, just the American government. He offered me one more cup of tea and I left.

This was a somewhat common occurrence throughout the Middle East – people would tell me they hated my country while treating me as a guest of honor. In the case of the Bedouins, nomads in the desert, it makes some sense – a traveler will be welcomed in and their needs met, even if they are an enemy.

In each country, people were eager to tell me their stories, as if to demonstrate that theirs was a reasonable course of action. The Israeli Jews, the Israeli Muslims, the Palestinians, the Syrians, the Lebanese Christians, the Lebanese Muslims, the Turks that at the time were invading Northern Iraq, the Kosovars who had gotten their own country only a month earlier, the Croats, the Bosnians, the Serbs. In each case, I was usually impressed by the validity and consistency of their story, and by the extreme paradoxes that emerged when trying for reconciliation, just as with $1 + 1 = 1$ and $1 + 1 = 2$.

Finally, now, I'd like to say a few things about the one thing that mathematicians do that is, or at least can be, a direct exercise of compassion: teaching. Teaching is a beautiful opportunity to show love and compassion. Every minute you spend trying to understand your students and help them learn is a minute that your heart is open and pure. It is a challenge, with consequences that can touch you as deeply as you will allow.

I think that to be a good teacher you must be a good person. You must know yourself and know how to quiet yourself, pay attention, and listen. The best learning comes from this space of openness. On good days, a student and I can transcend the teacher/student relationship – at least for a short while – and learn from each other as two people in the world.

I'm just beginning to learn how to teach well. In my short career as a teaching assistant and instructor, the best words I've ever heard from a student came this past summer from a freshman named Claire. We'd been working through some problems during my office hour, and she started asking questions about life as a grad student. We got to talking about some of the traveling I had done, and she told me her story. She is a skilled pianist with perfect pitch, and a thoughtful young woman curious to see the world – but her parents are pressuring her to find a career in finance. I carefully offered my perspectives on independence, curiosity, and careers. Claire came back the next day to my office hour to talk more, and at the end of the hour she timidly said, "Luke, I think you've just changed my life." She's now made plans to travel to Swaziland next summer, and I've hired her to come and fix my out-of-tune piano.

During my Human Kindness Experiment, I discovered that one of the simplest and most profound ways we can help each other is by living our own truths – by living the best lives we can. Teaching is an excellent opportunity for this. Although I might be teaching calculus, I understand that – whether vocally or silently, directly or indirectly – I am also teaching everything I believe in. So in the classroom I try to live my life by example and be the best person I can. This, in turn, helps me, outside of the classroom, to continually be true to myself and grow as a person.

One of my tabla teachers in India once told me, after I had finished one of my daily lessons on his living room floor, "Do you know, Luke, what makes a student succeed? It has nothing to do with how hard the student practices or how many compositions he learns. All that matters is how much the teacher loves that student."

Math and Meditation: Why I Do Math.

My father, who does yoga every morning, must've taught me some yoga or meditation when I was too young to remember it, and planted the seeds for a lifelong journey of self-awareness. For as long as I can remember, I've been curious about my mind, my body, my awareness, and the relationship among them.

I remember the first book I ever bought with my own money. I was probably 12 or 13 years old. I snuck to the bookstore, bought a book called *Yoga for the Spiritual Muscles*, and went home and hid it under my bed. It was a typical introductory yoga book, with big pictures and descriptions of poses, along with explanations of their effects on the mind and spirit. For some reason I was ashamed of my purchase, and kept it like a dirty little secret, only taking it out to practice before bed or early mornings on the weekends.

As soon as I was in college and had the option, I started taking yoga classes, in the style of BKS Iyengar. I was supremely curious about these "spiritual" pursuits, but also supremely skeptical of anything stepping outside of Western science. After my sophomore year I took a year leave of absence and went to India, and have since returned again and again.

The thing I love about India is that personal growth and development are very much a priority among the population. Here in America, in general, our priorities are centered more on material wealth and success. In India, I met more people who were willing to work only as much as needed to support themselves and their families, and not much more. Many Indians I met followed a spiritual teacher and openly discusses spiritual matters. Many regularly practiced yoga, which in its original form has a much wider scope than most Americanized versions.

In India I wandered the country and stayed in ashrams, encountering quite a range of "spiritual paths". For example, some people think that the way to enlightenment is to have lots of sex. Some people think that the way to enlightenment is to drink your own urine. None of the many paths really caught my fancy, but all were oriented towards developing awareness and understanding the mind.

I considered becoming a monk and studying my mind directly. Instead, I decided to study it indirectly, by studying mathematics.

People ask me why I do math. The answer is: I like the way it feels, I like the experience of doing it, and I like the way it allows me to watch how my mind works. I use math as a tool to understand my mind.

In May 2006, I did a Vipassana meditation retreat in New Zealand, which is a type of Buddhist meditation taught by S.N. Goenka. Mr. Goenka, originally from Myanmar, has developed a worldwide following over the last few decades. His approach is based on the original teachings of the Buddha, recorded in North India 2,500 years ago in the Pali language. Anyone curious and willing is welcome to participate in a free 10-day Vipassana meditation retreat, at one of their many centers around the world. You take a vow of silence, and for ten days you follow a strict routine of meditation, instructional periods, food and rest. In total you spend twelve hours each day sitting in meditation, and over the course of the ten days you learn and practice three meditation techniques. This experience had a profound effect on me in a way that other paths had not. Leaving the retreat, I was saturated with the overwhelming beauty of being alive, and carried few things. I carried these meditation techniques, which are tools for understanding our minds and the nature of reality. I carried a conviction that we exist to love and show compassion, and nothing is worth doing unless it is done with love.

In the fall of 2006 I started math grad school, and, as I mentioned before, it was quite a shock. It seemed that I had to demonstrate endurance and obedience for several years before I could apply my creativity and curiosity. We learned too much too fast to have any self-awareness of what was happening. As a result, the math was empty. It was like rushing along a trail in the woods and counting miles, rather than stopping to enjoy the views, flowers, smells, and the good feeling of fresh air and exercise. It was like driving over a mountain pass in the night.

I knew exactly why I was doing math: to learn about my mind, and maybe even the nature of reality as a mental construct. I would ask: if my reason for studying math is right here today, in the process of doing math, why do I need to run over there so fast? And anyway, if math is infinite, what is over there that isn't over here?

As I described in *Math and Nomadism*, in the summer after my first year, while studying for my first qualifying exams, I was able to reconnect with the depth that I knew was possible in the mathematical experience. I was free to structure my own studying, free to examine my process, and free to go slowly enough to watch my mind.

When I returned for the second year at school, I started on my path to being a "bad" mathematician. I was taking courses, but I would skip class and teach myself, and only do half the homework problems so I had time to think about them and think about thinking about them. I was busy asking questions that aren't usually asked. In math research, you take some system and ask some questions to try to get a better understanding of that system. The system I was looking at was the whole mathematical experience. I started with the assumption that "mathematics" has something to do with the humans that speak, write, create/discover, learn, and teach it. *How* do we do math? How do we do good math? How could we do better math? How could we connect mathematics to non-mathematics in other areas of life?

I started to seek out resources to answer these questions. Traditionally, mathematicians themselves don't ask these questions; they leave it to the philosophers and sociologists. But there are few philosophers and sociologists that have a good sense of what working mathematicians actually do, so little progress is being made. In the early 20th century, mathematics philosophy reached a stalemate between three schools – Platonism, Formalism, and Intuitionism. Very briefly, Platonism claims that mathematical ideas exist independent of humans in an absolute realm of Ideas; Formalism insists that mathematics is nothing but a body of formal manipulations of logic symbols; Intuitionism claims that truth is contingent on constructability, and insists on a new 'intuitionistic' logic. Without going into details, one cause of this stalemate was that they were all *a priori* philosophies of what mathematics is, and none really satisfactorily captured the experience of *doing* mathematics. Over the last few decades, however, some work has been done to develop a 'humanistic' philosophy of mathematics, which takes the practice of mathematics as its beginning.

Books started to collect on my desk: *The Mathematical Experience, Where Mathematics Comes From, How Mathematicians Think, Social Constructivism as a Philosophy of Mathematics, Eighteen Unconventional Essays on the Nature of Mathematics*, etc. Was this mathematics? Many would argue that it isn't. All I knew was that this was what I was passionate about understanding. I was ready to begin my mathematical – or 'metamathematical'[2] – research, and decided to do it my own way.

In December that year, I flew to Tanzania. I spent three weeks there, on the beaches of Zanzibar, in the mountains walking village to village, and climbing Mt. Kilimanjaro. From there I flew to Dubai, then Egypt. As I described earlier, from there I traveled overland through the Middle East up to Turkey, and then through the Balkans and up to Vienna, Austria.

On the trip I brought several of these metamath books. I had some friends in Tel Aviv that I was going to stay with, and had Amazon.com deliver several more books to their house. While riding buses and trains or sitting in coffeeshops, I carefully read and outlined each book.

As I read, I decided the next step was to start the conversation within the grad department in Seattle, to try to establish a community of mathematicians interested in investigating the personal side of math. While I was away I began to organize a reading group centered on these texts. Although I had the support of one faculty member, the department decided that this was officially outside the domain of mathematics, and thus couldn't be offered for credit.

In the spring, back in Seattle, the conversations began. We had six meetings and covered a few books and several articles; weekly attendance was between four and fifteen, including one faculty member, one librarian, and grad students of varying years. The dialogues were at times inspiring, at times surprising, and at times frustrating. In the end we

---

[2] The term metamathematics has already been given a meaning. In the 1950's there was an attempt to use self-reference to construct a philosophy of mathematics that was logically contained within the true statements of mathematics itself, and this unsuccessful field was called metamathematics.

stopped because everyone was overwhelmed with their other work.

I was disappointed that the reading group ended without any future momentum. But those dialogues had been productive for me. I had seen how at least a few mathematicians tried to answer these questions. As happens when you open to other perspectives, I realized that there was actually quite a range of positions on issues I had discarded as settled or moot – issues like Platonism vs. Formalism. I also realized that my own positions and arguments were quite inchoate. The next step, I decided, was to work to better formulate my own ideas and articulate what I was trying to say.

If a community of metamathematicians wasn't going to spring up easily, then at least I could strive to do my own math the way I wanted. Towards this end, I established a math process journal. Every day in the evening I would look back at the math I had done that day, and one by one try to carefully describe the experiences. In this way I tried to develop my mathematical self-awareness.

All this was informed by something unexpected that had happened during my trip to the Middle East, something that was going to have all sorts of surprising consequences. In the southern deserts of Israel in January, I sat for another ten day silent Vipassana meditation retreat. This time, on finishing, I promised myself I would maintain a daily practice. Every day since then, with few exceptions, I've meditated for one or two hours. I maintained this practice in spite of all the chaos of traveling. Coming back to Seattle, I maintained this discipline in spite of feeling the pressure from schoolwork. This practice has really changed everything.

I want to attempt to explain the meditation, and the effects it has had on my mind and my mathematics. First, though, I want to add a disclaimer. I'm not trying to teach the techniques that I learned. There's a reason that these techniques are taught the way they are, by a master during a ten day silent retreat. If you are intrigued by this meditation, I encourage you to go to dhamma.org and read more, and find a Vipassana center near you. Reading words describing the results of meditation is nothing like practicing meditation. It is one thing to understand an idea like "everything that arises will pass away" on an intellectual level, but through meditation it is possible to understand on a much deeper, experiential level.

The first meditation technique is called anapana, and the goal of anapana is to strengthen and focus your awareness. You are learning how to use your mind to study your mind, and to do this you must hone the instrument of awareness – just like in a science experiment, when you use tools to understand the inner workings of a system. This is done by focusing your awareness on a single object, your natural breathing. Without changing it in any way, you try to observe your breath simply as it is, not as you would like it to be or think it should be. You are not looking for any particular experience, merely paying attention to what is happening here and now. As you observe your breath, your mind will wander. Once you realize your mind has wandered, you simply return your awareness to your breath, without generating any negative reaction. Over time, perhaps, your mind will wander less, and you will become quicker to recognize when it has. The periods during which your awareness is strong and focused purely on your breath will lengthen. In this way, you develop a strong, unwavering awareness, as well as a meta-awareness that is able to verify that your awareness has not moved. You also learn to distinguish between the sensations you are experiencing and the interpretation and processing of those sensations.

In the second technique, called vipassana, you use this new tool of sharp awareness and direct it towards your mind and bodily sensations. According to Buddhist philosophy, the origin of suffering is in the link we develop between sensations and our reactions to those sensations – we are continually chasing after positive states and running away from negative states. In vipassana, you learn to break this link, to observe whatever sensations arise without reacting. It turns out that whatever arises will, after some time, pass away. Rather than develop craving and aversion, you simply sit with whatever experiences arise, use your focused awareness to feel them and understand them as deeply as possible, and then watch them pass. Over time, perhaps, you develop equanimity, which is a sort of balance of the mind that allows you to live each moment deeply while maintaining clarity and perspective.

What have been the effects of my meditation practice? Generally, the anapana practice has resulted in an intense increase in my concentration and ability to focus. It's more common that when I start working on mathematics, I become immediately absorbed into it; the sights and sounds around me are eclipsed. I can work for hours without feeling any fatigue or hunger, or the slightest distraction. If there's music playing, I don't hear it; if there are people around me I don't notice. Not only is my mind sharp and steady, but this sustained attention is of the highest quality. It's easier for me to think clearly and deeply. I focus my mind on a problem or idea and keep it there until the secrets reveal themselves.

It has amazed me how much more efficiently I operate mathematically. Spending an hour and a half in meditation every day seems like a lot, but I usually end up feeling like I've saved time, that there is more time in the day to do things. What would've taken me four hours instead takes me three or two.

The practice of vipassana is slowly developing my patience and balance of mind. I'm more able to be passionate about the work I do while remaining detached from the outcome. Math can be extremely frustrating, but it is the frustrating moments that have within them the most potential for growth. By patiently sitting with my exasperation, I've occasionally been able to persevere and break through to deeper understanding.

Most importantly, these techniques have allowed me to be more self-aware as I do math. While I'm working on math, absorbed in a topic, I'm also able to observe my awareness as it moves and my emotions as they arise and pass. I'm more able to do math and think about doing math at the same time. Through meditation, you can develop some stratification within awareness. There is the awareness that connects you to a sensation, idea, or experience. There is also a meta-awareness "above" this that observes your awareness – how it is moving, its quality and strength – and can decide to control or direct your awareness.

This layering of awareness, besides bringing much depth and beauty to my experience of mathematics, has also made me a better self-directed student. When you are teaching yourself something, you must establish within yourself both a teacher and a student. The teacher within you must guide and observe the student. The student must be allowed to become immersed in a subject and brought to the challenging edge of understanding where the most learning takes place. The student must be made to struggle, and sometimes fail. The teacher must quietly observe this, without interfering, and then step in to bring clarity and perspective to the process.

Given the conflicts between math and compassion that I discussed earlier, it's not at all clear that doing mathematics is the best way to become a bodhisattva – an enlightened "warrior of compassion". But my attempts to understand my mind through the mathematical experience seem to be bearing fruit. A mathematical lifestyle gives me the freedom to contemplate.

Part II: Four Tangents

Proof and North Indian Classical Music

North Indian Classical Music (NICM) is one of the richest and most complex systems of music on the planet. Its oldest roots trace back four millennia to the religious music of the Vedic era, but its most significant features come from the court musicians of the Mughal period, starting in the 1500's.

In a nutshell, all classical performances entail an exposition of a particular raga and a particular tala. Ragas are melodic systems, manifested by a voice or instrument such as the sitar, sarod, or sarangi. In the West most music is in either a major scale or a minor scale, which can be thought of as a choice of 7 tones from the 12 tones between octaves. In NICM there are supposedly 50,000 ragas, although only about 150 are performed regularly. There are 22 different tones in an octave, and a raga includes a fixed ascending and (usually different) descending scale of 5 to 7 of these. There are rules as to which of these notes are more important than others, which should be emphasized and which almost avoided. There are rules about particular embellishments when ascending or descending the scale, and certain characteristic phrases that must be played. Each raga also has a particular time of day during which it should be played, and a particular season of the year. Each raga is defined by these various characteristics, which have become codified over hundreds of years. But most importantly, to each raga is associated a rasa, or emotion (literally, "taste"). There is an ancient Indian theory of rasa that goes back millennia and pervades all Indian arts. There are nine main rasas – lust, comic, pathetic, furious, heroic, horror, odious, wonder, and tranquility.

A tala is a rhythmic system with a similar multi-leveled structure, classically played on a pair of drums called tabla.

For example, rhupaktal is a seven beat cycle, broken into 3+4. The first three is called khali, the second four is bhari. Bhari is subdivided into 2+2. There are certain rules about what must happen or can happen during khali and during bhari. There is a specific base beat or theka for each tala, which in this case is "Tin tin na, din na, din na". Every tabla composition begins and ends decisively on the sam, or one, just as every melodic composition begins and ends on the sa, or tonic.

Traditionally, the goal of a NICM performance is to evoke the raga's associated emotion. This is done by a careful exposition of the raga's properties. This exposition also has a very particular development, starting with slow arrhythmic meditations on each note of the scale, ascending and exploring the structure over a few octaves. After this, the tabla joins in with rhythm, and a few chosen compositions are used as a basis for improvisations within the raga's framework. Later, the tempo accelerates and the music transitions to rhythmic play between the instrumentalist and the drummer, ending in a climax on the sam. Performances typically last around 90 minutes.

This rich structure of melody and rhythm is very engrossing. As you attend concerts, you slowly develop an ear to understand the languages of raga and tala, and at each performance you can hear a little more of what is going on, what is being communicated by the musicians.

Interestingly, in NICM the audience plays an integral role. Whereas in the West a performer is blinded by stage lights and the audience is silent and in the dark, at a NICM concert the house lights are left on and the stage lights are low, so the musicians can look into the eyes of the audience. The audience will make particular gestures and sounds to demonstrate appreciation throughout the performance, although clapping at the end has also been imported. I've been able to watch how different audiences affect a musician, and how the skilled musicians alter their performances to match their audiences. The concert really is a shared communication.

The Indians like their classical music all night long, and in large doses. Twice I've been to the Dover Lane Music Festival, in Kolkata in January. This is a week of all-night concerts, starting at 7pm or so and going until sunrise. It's hard to describe the state this puts you in – sleeping during the day and spending your nights at the feet of the masters of raga and tala.

During my first trip to India, I was a bit of a music junkie – going from festival to festival, in Kolkata, Delhi, Gwalior, Pune, Mumbai, and Varanasi. When I went back to India in 2004 I did the same thing. I have been fortunate enough to see the most famous Pandits and Ustads, but I've learned so much about music and communication from the novices as well. My years of tabla study, and a few months of classical vocal music study, helped my ability to access the music.

I said that the goal of a NICM performance is to evoke the raga's characteristic rasa, but this is not quite true. During the performance of, say, Rag Malkauns, you may be brought to tears and filled with sorrow. But the most beautiful thing is when the music ends and this strong emotion leaves you, and you realize you've been filled and emptied like a cup. You walk away with a transcendent sort of empty bliss, floating on contentment and gratitude. When this happens, it means that you and the musicians truly affected one another.

There are many close parallels between NICM and mathematics. The central object in the doing of mathematics is the proof, and in my mind a proof is a narrative very similar to a NICM performance.

For a while it was thought that proofs represented some form of absolute knowledge. The Formalists built their

71

philosophy on the idea that all mathematics could be reduced to a series of fundamental logic steps. They even undertook to demonstrate this. The famous example is Russell and Whitehead's *Principia Mathematica*, published in 1913, in which it takes 379 pages to prove that $1 + 1 = 2$. This is a quite unsatisfying definition for proof. No one would ever take the time to reduce every step to basic set theoretical logic deduction. Maybe a computer could, in theory, encode every theorem this way, but who wants to trust computers with the burden of absolute knowledge? Why not talk about proofs as they are actually created and used by human mathematicians?

The goal of every proof is to convince its reader that something is true and, ideally, to give some idea as to *why* it might be true. This can be done in many different ways and depends on the audience. For an undergraduate course you might spend a whole class period giving a careful and detailed proof; for a graduate class you might only have to say "Use Whitney's Embedding Theorem and apply a partition-of-unity argument". There is a wide range of established proof techniques and tricks, and the majority of a proof is a layering of combinations and applications of these using more or less cleverness. There are usually one or two key ideas around which the proof hinges. Occasionally there is a truly novel step, which will be recognized and added to the reader's proof-making toolbox.

There is a great degree of subjectivity in how a mathematician may lead you to the intended conclusion. Too much detail often obscures the key ideas, but nothing is more frustrating than a string of "It clearly follows that..." statements. Only the best proofs will help you to recognize the key ideas, which is unfortunate because this recognition often suggests new directions. The path to the conclusion is a very personal one, based on jumps of intuition and comprehension. The proof is an attempt to retrace this journey in a universal, almost linear way, so that it induces a sequence of "Aha" moments in the reader that suffice to convince and illuminate.

In a larger sense, the goal of a proof is not just to convince the reader. It is to leave the reader with a greater understanding of the material at hand, a sense of satisfaction, joy, and maybe even wonder at the austere beauty of mathematics.

Both artists – the mathematician constructing a proof and the classical Indian musician performing a raga – are extremely restricted by rules of what they can and cannot include in their creation. Within these rules, both must use a high level of intuition and artistry to weave their story, leading the audience very carefully. The challenges are formidable and almost paradoxical: for the musician, he or she must use the prescribed language of the raga to evoke a particular emotion in the listener; for the mathematician, impersonal deductions must suffice to induce a personal understanding in each individual reader. If the music or proof are successful, then once they finish we are immediately impressed by the larger purpose of the art – that sublime satisfaction and gratitude at having been taken on a journey through our own minds and feelings.

I like this broader, deeper metaphor for proof, because it seems to capture what proof-making is actually about. It is not linear or formulaic, but rather a subjective, personal story that we tell to each other to convince and illuminate.

North India Classical Music and mathematics are both rich theoretical structures that have developed over millennia. Both have developed a powerful language that easily allows deep communication. Before a NICM performance, the vocalist or instrumentalist will often meet the tabla player only a few minutes before they walk on stage. All they need to agree on is what raga and what tala to perform. Likewise, it's common for mathematicians to travel around the world giving detailed talks on abstruse topics. Despite all the cultural differences, we can always be perfectly clear with our mathematics. In both musical and mathematical dialogue, there is a constructive overlapping of realities from which emerge new shared ideas and expressions.

There is much discussion among today's classical Indian masters about the evolution of their art, about purity versus change as musical cultures around the world continue to meet and mix. If a tradition resists change, it will eventually become antiquated and irrelevant, and die. If it changes too quickly and broadly, it will sacrifice its integrity and cultural significance. Over time the tradition must adapt, preserving some elements while changing to include new influences. In the same way, we can look at the future evolution of mathematics, and wonder what familiar elements it will preserve and what revolutionary new ideas it will grow to include.

Algebra, Chaos, and Love

Clouds reflect the human condition. As long as humans exist, we will always strive to discern patterns in the world around us. To stare at the clouds is to watch our desire for understanding trace across frivolity. Some people decide they see a face, or a hand, or a man walking a dog. These are the patterns and shapes we are familiar with, but surely our minds are able to generate and recognize patterns that are deeper, and more abstract. We can, and must, stretch our imaginations to create new – truly new – structures and patterns.

An excellent example comes from chaos theory. Think of a dripping faucet. Sometimes the dripping is quite regular, but under certain conditions the dripping seems completely random. In this state of "chaos", the smallest perturbations in water flow in the pipe will grow exponentially and affect when the faucet drips. The classical scientific approach using fluid dynamics and differential equations will in practice be useless at predicting the dripping pattern. According to the classical physicist, the water is, of course, following the laws of physics – the patterns that all matter moves within – but the system is so complicated that we have no hope of seeing the pattern within the chaos.

Chaos theorists were able to listen to the dripping faucet and find a totally different, very deep and very beautiful pattern. Imagine recording the time each drip happens starting at a fixed time, and compiling the numbers in a list t_1, t_2, t_3, t_4,... Now make a new list that keeps track of the time between drips, d_1, d_2, d_3, d_4,..., where each d_i is the difference $t_{i+1} - t_i$.

Now take consecutive pairs of points in this list, and create a list of pairs, (d_1,d_2), then (d_2, d_3), then (d_3, d_4), etc. So far, all we've done is take a seemingly random list of numbers and make another seemingly random list of pairs. Each of these pairs can be thought of as a point (x,y) and plotted as a point on a two-dimensional graph. Now we have a sequence of seemingly random points on a piece of graph paper. There's no reason to suspect that our picture will show anything but randomness. Indeed, although the faucet is governed by the laws of physics, those laws don't make any predictions of what picture might show up on our graph.

As James Gleick describes in *Chaos*, when early chaos theorists did this they found that the points jumped around the graph, but after plotting many, many points a definite shape started to emerge. Given a point, there was no easy way to predict where the next point would show up, but there was regularity in the probabilities of where the points landed. This nebulous shape is called a "strange attractor", and is a pattern in the chaos of the dripping faucet. The strange attractor exists in this "phase space" of the system – a two-dimensional graph conceptually removed from the physical system of pipe and water. By looking at the faucet data in an imaginative way, scientists were able to see a new, beautiful type of pattern.

I grew up playing in the woods. My family lived down a long dirt road in rural Upstate New York. The backyard was a large grassy meadow that met a creek, and behind the creek was a forest. This forest was where I spent most of my childhood.

That forest was infinite, as far as I could tell. On brave days my neighbor James and I would venture straight back as far as we dared; we would find swamps, stands of cedars for making forts, small clearings where the deer slept, and endless other surprises. But after a point we were scared to go further, and would return to the house.

I think this is why I grew up believing in infinity, and believing in the infinite potential of imagination.

Once I got to college and realized that math was as varied and bizarre as it is, I concluded that math was infinite. Math was an expression of pure imagination, accompanied with some level of rigor. Any conceptual structure you could imagine, if you could describe it to others, could become math. Math was art with ideas.

It turns out this is not quite true. In order for something to be recognized as mathematics, it must be couched in a certain language and must fit into a history. The shape of math is in one sense like a tree; everything must build on something before it. No matter how bizarre it gets, it must be traceable back to the roots in a finite number of small steps. Math proceeds in very small steps. In some ways, the development of math is quite primitive; the basic notions of space and number are compounded and abstracted over and over again. Most math uses the same tricks and techniques to build on what has already been done.

Algebra is the study of mathematical structures. The structures that are studied in algebra provide the frameworks in which other areas of math can be developed. In this way, algebra is the closest to my dream of endlessly creating newer and richer patterns to use in comprehending the world.

How does math progress? Different types of math develop in slightly different ways, and here I'm only going to talk about the way that algebra progresses. First, start with a system and work strictly within this system. Try to understand this system as completely as possible. This in itself will often lead to many questions, some more difficult than others. Once this system is relatively well-understood, go outside of it. Break some rule of the system, or add a new rule, or compare this to another system. In this way, you enlarge the system under investigation to encompass new territory. In most cases, the original system is now seen as a special case or specific example in some larger framework. The new system transcends but includes the original, and now you repeat the process.

At each transcendent step, there is some new idea. There is a creative act that generates a new, really new idea, and allows for the breakthrough to the larger context. Sometimes the idea is clear and monumental, other times it is tenuous and must be nurtured. Sometimes there are several ideas, several ways in which the system can be expanded; in these cases usually the most interesting one is chosen first. In each case, after the idea has been created and a new system invented, the new possibilities must be explored.

The classic example of this process is the development of our various number systems. At first there were only the counting numbers: 1,2,3,.... The invention of zero – the something that is nothing – was a monumental breakthrough. Then the question was posed, "When two numbers are added, the answer is another number. Is there a number such that when added to 5 the answer is the number zero?" In modern terms, this was asking for a solution to the equation $x + 5 = 0$. A new breakthrough was required, to expand the notion of number to include negative numbers – thus giving us the integers. In the same way, the rational numbers, which are all the fractions and include the integers as a special case, were soon required for solutions to equations like $2x = 3$.

The step from rational numbers to real numbers was another very upsetting breakthrough. Given a square with side of length one, how long is the diagonal? The Greeks believed that every length could be correlated with a number, and so this length, which is the square root of two, should be a number. But the square root of two is not a rational number. Supposedly, when a member of Pythagoras' society proved this, he was killed in order to keep the result a secret. Eventually, though, there was no denying it, and the meaning of number had to be enlarged to include all the real numbers.

By many accounts, the real numbers can hardly be said to be completely understood. But looking for a solution to the equation $x^2 + 1 = 0$ has led to the creation of the complex numbers, which transcend and include the real numbers.

I want to compare this process to what it means to me to love someone. I think there are two main components to love.

The first is complete, unwavering acceptance and support. To love is to embrace the truth as it is, without judgment and without questioning. You must strive to understand as deeply as possible, and long for unity and communion.

This is somewhat analogous to working within a mathematical system, embracing the arbitrariness and trying to build as complete an understanding as possible.

The second component to love is a yearning for something more, a longing for transcendence. You must always challenge and push your partners. Only you can set yourself free, but someone who loves you will try to help you in your continual evolution towards freedom. The point is not to reach any particular destination, the point is to be continually growing and seeking. Together you can establish a space for breakthroughs to occur. When they do, you nurture them and explore the new possibilities that have been created.

This is analogous to the transcendent step in the progress of math. A creative breakthrough enlarges your world.

These two dimensions are clearly contradictory. You can't be completely accepting while also pushing someone to change and grow. You can't work within a system while breaking its rules at the same time. There must be a balance between the two.

Too much of the first component will lead to stagnation and dependence. When a couple becomes complacent and stops trying to change, it becomes stuck. An area of math without enough breakthroughs and advancements becomes boring and gets left behind.

Too much of the second component will lead to chaos or dissolution. Without a healthy balance of support and acceptance, pushing and challenging will be unproductive and harmful. It's not possible to move towards transcendence without a strong connection based on understanding. Likewise, generalizing mathematics must be done in conscious steps; too much abstraction without a solid grounding in intuition will be hopeless.

I identify as bisexual, and my approach to love has been shaped by the fact that no single person can complete me. Rather than move towards a conventional monogamous relationship, I have creatively explored the wide range of relationships and manifestations of love. Often these are open and involve multiple partners, and thereby require an incredible amount of honesty, self-awareness, sensitivity, and presence. Once out of the societal relationship norms, which for me are too heavily weighted towards security and stability, my partners and I are more free to explore and create unique connections.

This exploring and creating proceeds with love, and follows the pattern I've described above, coupling acceptance and support with challenge, independence, and growth. It is a testimony to imagination and the belief in infinity. Like the space of the sky and its clouds, the people in my life create a landscape of emotions and thoughts. Together we explore that landscape, seeking to illuminate the beautiful patterns and forms within it.

Feminism, Math, and Abstract Nonsense

Sir Francis Bacon was an English philosopher in the sixteenth and seventeen centuries who played a key role in the Scientific Revolution. He was the first to clearly formulate the Scientific Method – the inductive process of performing experiments, drawing conclusions, and conducting further experiments to test new hypotheses. This was the beginning of science as we know it, the process of building an understanding of nature. He was also attorney general for King James I, and at the time this meant conducting frequent witch trials. It was not uncommon for mechanical torture devices to be used during these trials. Metaphors from the courtroom entered his scientific writings:

> Nature, in his view, had to be "hounded in her wanderings," "bound into service," and made a "slave." She was to be "put in constraint," and the aim of the scientist was to "torture nature's secrets from her.[3]

Does this say anything about science as it has developed over the last four hundred years?

My literature of escapism as a twelve year old was popular science, in particular relativity and quantum theory. I remember reading interpretations of the bizarre philosophical implications of relativity and quantum mechanics, in books like Gary Zukav's *The Dancing Wu-Li Masters* and Fritjof Capra's *The Tao of Physics*. There are many great books written on this "new physics" and what it might mean for the world as we

---

[3] Fritjof Capra, *The Turning Point*, 1982. p. 56.

know it – some very Eastern, some very Western, some very grounded in the science, and some totally ridiculous. They talk about many new ideas: the Heisenberg Uncertainty Principle, which fundamentally restricts the amount of information we can simultaneously have about, for example, a particle's position and speed. In all interactions, there is an unavoidable interconnection of a system and its observer. Every particle is both and neither a particle and a wave, depending on how you look at it. Matter is energy and energy is matter, and both fundamentally have a grainy or "quantum" nature. Energy stretches time and space, which are together a single four-dimensional fabric.

The reason this physics was called "new" was that it presented an entirely different paradigm for nature and our relationship to it. Classical physics and most science, descended from Sir Francis Bacon and the other "fathers" of science – Descartes, Newton, etc – thought of nature as a machine to be broken apart into pieces. It was to be understood so it could be controlled and exploited. The new physics challenged that view by denying determinism. It showed empirically and mathematically that things are not so linear.

My fascination with this physics and its implications continued into college. In the summer after my freshman year I worked on an organic farm in New Jersey. The people there introduced me to complexity theory and its own New Age interpretations. Complexity theory studies whole systems and their dynamics. Rather than break a system apart, and focus only on those systems simple enough to be understood in their fragments, it looks at large and complex systems – the weather, populations, the environment, the stock markets. It attempts to identify holistic behaviors and the transition between global states. In complexity theory you attempt to characterize emergent patterns, the beautiful dance on the edge between chaos and order. You describe exactly how the whole is created and comes to be more than the sum of the parts. Closely related fields are general systems theory, nonlinear dynamics, and chaos theory.

The claim has been made that this presents a more "female" type of science[4]. Early science was full of yang – masculine, aggressive, competitive, rational, analytic. Complexity theory was the beginning of a return to yin – feminine, responsive, cooperative, intuitive, synthesizing.

I read lots of books to this effect, and perhaps I was brainwashed. Was there really a paradigm shift at hand? Were we entering a new era, in which we would cooperate and integrate in order to feel one with nature and at home in the universe? I wanted so badly to believe it.

There were a few setbacks to the movement towards a holistic, non-reductionist science. One was the invention of computers and the accompanying way of thought. Using the computer as a metaphor, we conclude that life is nothing but a collection of programs and sub-programs and sub-sub-programs. There are ones and zeros, and it's tempting to believe that everything can be reduced to a collection of ones and zeros. The way to understand something is still to cut it into smaller and smaller pieces and see how they fit together.

Another major setback was the mapping of the human genome, which seemed to confirm that life is nothing but sequences of Cs, As, Gs, and Ts. This was an incredible accomplishment, but philosophically it tells the story of reductionism and determinism. Depressed? Maybe it's in your genes somewhere. How will you die? It could be written in your genes.

I almost ended up as a complexity theorist. After my fourth year of college I applied for a summer internship at the

---

[4] For example, see Fritjof Capra's *The Turning Point*, 1982, or Joanna Macy's *World as Lover, World as Self*, 1991.

Santa Fe Institute. SFI is one of the only places in the world where they actively study complexity. They have a continually rotating faculty from a wide range of fields – computer science, economics, biology, psychology, sociology, and more. They bring these people together and ask them to look at the big picture, to talk about the patterns that transcend the academic divisions we've created. I didn't get the internship, and ended up spending the summer doing research with lasers at the Lawrence Berkeley National Lab in Berkeley, CA. It was an excellent summer (I bought a motorcycle...), but in the end I decided that physics wasn't for me. The people there did physics like it was their job, not like it was part of an eternal human quest for realization and discovery. There wasn't enough imagination. I decided I would (eventually) go to math grad school.

If science has some masculine-leaning imbalance to it, surely mathematics is worse. All the objects in mathematics must be pinned down and dissected dispassionately. We learn math so that we have mastery over the ideas. Never mind that they were created in our minds and are sustained in our thoughts and conversations – the math is out there existing objectively, and we're in here looking at it and poking at it. Mathematics is so intricate and complicated that we must chop it into pieces and study its fragments; we must specialize to the point that only a handful of people on the planet can understand our work. And if there is any personal, human dimension to what mathematics is, we sweep this under the rug and avoid acknowledging it as much as possible. We label it intuition, a dangerous and deceptive temptress leading us away from reason.

I'm not sure if this extremely yang energy of mathematics, and its reductionism and determinism, have anything to do with the fact of the gender imbalance in mathematics. This would be claiming that women act

84

according to feminine characteristics and men act according to masculine characteristics, and I'm not prepared to do that. But sometimes being a mathematician does feel like being in a boys-only club. It's hard to imagine that the predominance of males in mathematics has not affected mathematical philosophy and practice.

At the end of my first year in grad school, I had a meeting with the Graduate Program Coordinator. We were sitting in his office on the top floor of the math building, a moderate sized room filled with shelves of books and stacks of papers, looking just as you'd expect a math professor's office to look. I said to him, "Suppose that mathematics was all of this room. Now, after four years of college math and one year of intense graduate school, how much math do I know?" He thought for a while, then made an invisible ball with his hands about the size of a grapefruit, and said, "I'd say you might know about this much. No, maybe a little less."

The point was that there's a lot of mathematics – too much for anyone to really get a sense of it all. The context was that he was telling me it was time to specialize, time to find an advisor and become an expert in a tiny little piece of the entirety of math. In that moment I saw a tiny little thread winding its way out from the grapefruit of my core math knowledge, to some arbitrary little ball, maybe the size of a pea, in the volume of the room. This, I feared, was the shape of math. By devoting my life to math, I could look forward to that pea growing to the size of a plum. I was discouraged. In the next instant I saw a flash of something different; I saw the volume of the office filled with a beautiful web. I saw a strange network, with a strange architecture to it, connecting all the areas of the room into one whole object. I thought: this is what math *could* look like.

After all, isn't it up to us to decide? We create math and actively participate in its evolution; we can decide what kind of math to create. The question is not just what can we do, but what should we do? What should math look like? What is the most beautiful architecture we could give to the whole shape of mathematics?

Then I began to wonder, if we implemented a more holistic or more feminine mathematics, what would it look like? In concrete terms, what would this mean for the practice of mathematics?

In a "feminist" mathematics we might try to account for the personal side of math. Rather than hide it and ignore it, we might embrace the power of intuition and the central role it plays in everything we do. We might study mathematics as a human endeavor. This whole book, in a way, is an exploration of this idea and small attempt to do this.

A holistic mathematics would sacrifice depth for breadth more often. We will always need the specialists, but to balance the yin and the yang we need people devoted to integrating and synthesizing, to looking for the larger patterns and gestures of mathematics as we know it now.

This is not a hopeless ideal, and in fact math is not as fragmented as I've made it sound. It turns out that there are structures in mathematics that unite and interconnect across large conceptual distances. Generally they are called isomorphisms, and they appear in all areas of math. In physics, $E=mc^2$ is considered a beautiful equation because it equates two things that were once thought to be completely different – energy and matter. Similarly, in mathematics there are countless isomorphisms between mathematical objects, many of which on the face of it are completely different.

What's better yet, there is an area of mathematics dedicated to generalizing the idea of an isomorphism to give an overarching framework in which all of mathematics may be framed. It's called category theory, and it was invented in the 1950's. Category theory gives a way of translating between fields of math. For example, a problem in algebra can, using category theory, get pushed over into the geometry domain in which it is a different, geometric question. We can then use the machinery of geometry, and geometric intuition, to reason through a solution, and push the solution back to see the implications in algebra. In this way, what may be a seemingly impossible problem in one area can occasionally be translated into a tractable problem in another. After the process, we have a somewhat enlarged sense of what's going on; these objects are not algebraic objects or geometric objects, but some sort of algebro-geometric object that can be projected into the two domains.

Category theory offers a first step in integrating mathematics. But there are no classes on category theory itself. It is only first introduced, in small doses and always in the context of particular applications, in the first year of grad school.

Think of how small a slice of the population has seen this holistic, integrative approach to knowledge! Might there be implications to the philosophy of mathematics, or the philosophy of science, or generally to the way we engage our environment every day?

Most mathematicians think of category theory as hopelessly abstract. They complain that although it is a powerful and useful tool, you never really understand what it means. They're used to the kind of understanding that comes from focusing on a small field – an intuition built on familiarity and experience. Category theory seems too general to formulate this kind of intuition, and so when used can feel dissatisfying. Often arguments using category theory are referred to amiably by mathematicians as "abstract nonsense".

In fact, as of this writing, if you enter the phrase "abstract nonsense" into Wikipedia, you will be sent to an article on category theory!

This lack of natural intuition of "what's really going on" in a category-theoretical argument is not a valid reason to dismiss it, I believe. No one will argue that category theory is not hard and abstract. But have you ever tried to see in four dimensions? A one-dimensional sphere is a circle, and lives in the two dimensions of a plane. A two-dimensional sphere is the surface of a ball, living in three dimensions. What does a three-dimensional sphere, living in four dimensions, look like?

Biologists and physicists will tell you we can only see three dimensions. But if you ever ask a geometer about a four-dimensional object, like a 3-sphere embedded into 4-space, you may get the impression that they can actually see it in their mind's eye. Intuition is developed from knowledge and experience. Some mathematicians have built a deep understanding of how a 3-sphere behaves; they have such an intuition about the object that they might even claim they can see it.

So, while mathematicians tend to be skeptical, unappreciative, or simply unaware of their intuitions, we often implicitly use intuition as a qualitative gauge of our level of understanding in an area. I see intuition as a central goal of mathematics; to understand without having developed an intuitive sense is to not know.

I wonder if it isn't possible, through experience and familiarity, to develop a natural intuition about category theory and what it means about math as a whole. This is the kind of math I want to do. I want to try and find out if it's possible to see the entire, beautiful, integrated shape of mathematics.

Math, Rock Climbing, and Existential Crises

Recently I hiked a section of the Pacific Crest Trail, and ended up staying at a hiker hostel run by a wonderfully generous old couple. They had a strange ritual they performed every night before bed. Dressed in their pajamas, they would look at each other and he would say to her in animated disbelief, "This is it?" She would respond, "This is it, buster!" They'd do it three or four times, back and forth, until they were convinced with their performance, then they'd go to sleep. It seemed both absurd and wise at the same time.

I think everyone is presented with existential dilemmas at various points throughout life. I *hope* everyone is. We all devote so much time and energy to meaningless and arbitrary puzzles. What's the point? Why bother doing this, or that, or anything?

There is no real solution. You can decide to kill yourself, or you can push it out of your mind until it returns. It always returns, because no matter how hard you try to forget the absurdity of life, death will remind you.

Most peoples' solution seems to be: fill the emptiness to avoid the emptiness. Keep busy with the games of life – games of tension and release of varying scope and complexity. So we enroll in a school, or commit to a relationship or a job. We pick something among the arbitrary choices, and we make the commitment to stick with it. We convince ourselves it matters.

Mathematics is merciless with respect to existential crises. Why should anyone be so passionate about something so blatantly absurd and imaginary? Unless you slide over to Applied Mathematics, all your options for study – Combinatorics, Number Theory, Algebraic Topology,

Analysis, etc – are equally absurd. There's no hiding by believing that math matters, because really nothing you do or could do matters when you're doing higher mathematics. As a mathematician, every day you're confronted with the dilemma, and every day you must embrace the arbitrariness. You must pick a problem that seems interesting and tractable to you and commit to it. Some math problems take years to solve. Some math problems take lifetimes to solve. Some math problems have taken the lifetimes of hundreds of mathematicians over centuries, most of who worked tirelessly and never saw the final fruit of their labor.

The situation of the mathematician reminds me of the situation of the rock climber. Rock climbing routes are rated for difficulty. No one is interested in climbing a route that is too hard for them; this is frustrating and futile. But no one is interested in climbing a route that is too easy for them; this is boring. So it's not just about getting to the top. Climbing is fun and worthwhile only if you climb in the range such that you are challenged. Maybe you succeed, maybe you don't. You try to find a route that will absorb your attention and efforts, you work at it until you get it or give up, and then you look for another route.

We are all at different places in our answer to the existential question. Of all the people I've met, it seems that a bus driver is just as likely as a monk to have found peace of mind. There are definitely mathematicians that glow with a cheerful serenity, seeming to relish every theorem and example as though therein lay the secret to happiness. There are also mathematicians with empty eyes that act like hollow machines, and it is not uncommon to abruptly take a leave of absence citing "personal reasons". I can't claim there is more of one type than the other, but I think the distribution may be more pronounced than for other walks of life.

Among the rock climbing community, I do think there are a greater number of truly content individuals. Rock climbing is a very ethically-oriented way of life. Anyone can climb anything with enough assistance and cheating; what matters is *how* a climb is done. Even after a climber makes it to the top, she or he may repeat the route again, and again, until it can be done more gracefully and effortlessly. These ethics extend to environmental stewardship, and the climbing culture is ingrained with values of environmental sustainability, simplicity, and appreciation.

Another prominent dimension to climbing is the community itself. Climbers are among the warmest, friendliest, and most generous people I've known. The sport necessitates camaraderie and demonstrates the positive aspects of codependence built on self-reliance. Climbers trust each other with their lives, and this builds strong social bonds.

In the face of an incredibly meaningless and arbitrary endeavor – pulling oneself up a rock and lowering back down – the rock climbing culture finds meaning in their values and in their community. I think the same can be said of mathematicians. If I spend my days chasing symbols around a chalkboard or discussing the convergence of a spectral sequence with my colleague, we find meaning in each other and in our shared mathematical experience.

Part III: Interior, Boundary, Exterior

Volcanoes, and Math as Religion

How does the math community look from the inside? It looks like a religion. Our community is based around a set of shared beliefs and common behaviors. Much of academia is like a religion. There are elaborate rituals centered around exams and progress in the knowledge hierarchy. There are robes and words intoned in languages that few understand. There are initiation rites, such as that intense first year of graduate school and the subsequent qualifying exams.

The power hierarchies in mathematics, and all academia, are based on knowledge. Lowest are the undergrads, then the young grad students, the older grad students, the postdocs, the non-tenure-track faculty, the tenure-track faculty, the tenured faculty, and finally the old and wise sages with offices on the top floor. Never have I been treated with such condescension as when I began as a grad student.

These levels of initiation and knowledge remind me of the Santeria religion common among the Afro-Caribbean communities in the Americas and West Africa. Cuban bata drumming is central to all religious ceremonies in Santeria. The drumming ensemble is three musicians each playing double-headed drums in rich polyrhythmic songs. Each song is associated with a particular orisha or spirit, and the music is used to channel the orisha and establish a communication with the divine.

  I attended a few Santeria ceremonies in Philadelphia. The participants surround the drummers, singing and dancing. Every so often, someone dancing near the drums will become possessed by the orisha's spirit, and begin to shake and flail.

He or she is shielded from the others for safety, and the experience is allowed to run its course. Eventually the spirit moves on, and the person is escorted, exhausted and weak, away from the drums to rest. It's quite intense to watch.

The priests of Santeria are the drummers. I remember, before I knew this, being introduced to Peachy, a wonderful, smiley old black man with incredible chops. I played conga with him a few times, accompanying some African dance classes. Later I found out that he is the High Priest among the Santeria community in Philadelphia.

Among the drummer/priests, there is an elaborate hierarchy. This hierarchy is centered around knowledge, specifically knowledge of rhythms. Every song has layers of secret rhythms associated with it. When I learned bata, we learned a secular version with all the sacred, secret beats and phrases removed. As a beginning Santeria drummer, you will be taught a few sacred beats. Over time, you will be taught a few more. As you progress up the hierarchy and prove yourself as a drummer/priest, you learn more and more secret beats. These allow you to more fully channel the orisha when you perform. Each step in the process is celebrated with its own initiation ritual.

The content of mathematics – all the definitions, axioms, and theorems – is our shared belief. It may seem strange to call mathematical knowledge belief, but that's all it is. Everything in math is built on various axioms, which must be taken on faith. It is built up using sequences of logic steps, the validity of which must also be taken on faith.

Could anyone question the axioms and the logic that are used to build math? Questioning axioms is actually central to the development of mathematics. Also, in some areas of math, nonstandard logics are used. Even the most "reasonable" logical principles are questioned occasionally. In the early 20th century, a philosophy of mathematics called Intuitionism denied the law of the excluded middle – that something is either true or not true. The intuitionists threw out most of mathematics as based on invalid reasoning, and attempted to start from scratch without this simple logic idea.

Faith, blind to various degrees, is central to mathematics. Math develops through proof. Proofs are how mathematicians establish new math. Proofs are also where subjectivity and faith enter mathematics. A successful proof must convince its reader that the conjectured statement is true; that is all. But every reader comes from a different place, and few written proofs are accessible to everyone. Proofs skip steps that are taken to be obvious by the author. When writing a proof, you always use other results, of whose truth you have somehow been convinced.

In practice, often you are convinced that something is true without fully understanding why it is. If all of the world's algebraic geometers agree that a proof is valid, then I'm convinced. If the top five most famous algebraic geometers in the world agree that proof is valid, that's enough to convince me. I think, "Mathematicians are pretty smart, and pretty careful; I'm not going to take the time to try to understand the proof." I may even quote the result in a proof of my own. If my proof is novel, then my faith in mathematics has been used to advance mathematics. These proofs that are the bricks and mortar of mathematics are as reliable as the tooth fairy. History is full of proofs with flaws, some of which weren't found for years.

Last year I took a course on Class Field Theory (whatever that is). The material is notoriously challenging, but there was no course textbook, only lectures. The professor had a habit of going too fast, and skipping proofs that he thought were uninteresting. In the proofs he did present, he would often sidestep large parts of the argument with, "It's obvious that...", "Clearly...", "I'll leave this as an exercise", or "You should look this up". I once read that the purpose of proofs in mathematical literature is to convince, and the purpose of proofs in mathematics lectures is to explain. This professor did neither.

As time went on, we were left with few choices. We could take it all on faith, and add these results to our body of mathematical truths. This faith would be based on faith in the authority of this particular professor and the words coming from his mouth, or on faith in whatever resources he was working out of. Or we could fill in all the details on our own, and convince ourselves of the things he assumed or omitted; in other words, we could teach ourselves the material.

I remember once during a lecture he said, "This is an isomorphism, and here is the map that gives the isomorphism." I decided to try to work it out on my own, to see that the map was indeed an isomorphism. I spent a few hours on it and made no progress; it seemed there was a mistake. I went to his office hours, and he confirmed that he had given the wrong map. For 45 minutes, we tried to figure out what map would work. Eventually he went online to the lecture notes he was working from, of another professor at another school. Yes, those notes had the same incorrect map, and my professor had simply copied the map without verifying it was correct. I gave up, and the professor worked on it alone for some time before emailing me (and the author of those notes) the correct map.

During this process, I was sure that there was some correct map that would work; I was sure that there had been a small mistake, but this was not a flaw that would cause Class Field Theory to crumble into inconsistency. But this sureness was simply a matter of faith. I believe the things that I've been taught. I believe that math is right. Mathematicians argue that this must be the case, since all new results are examined as carefully as possible by a collection of experts before they are published and thereby taken for true. But there is clearly room for error in this process. Furthermore, they argue that mathematics is continually re-verified as it is taught – the teacher and the student retrace the logic and confirm it. But my experience in Class Field Theory showed that this is not necessarily the case.

Eventually I stopped going to that class. I lost faith in the professor. He might present a theorem and sketch a faith-based proof, and I might write it in my notebook, but I was unable to say that I knew that theorem was true.

In the summer of 2005 I sailed on a cruising yacht from New Zealand to Japan. While in Japan, I decided to try to climb to the top of Mount Fuji. Most people who summit Mount Fuji do so by taking a bus two thirds of the way up, and then hiking four hours in the early morning to reach the top as the sun is rising. I imagine they take lots of pictures of that sunrise; it must be pretty spectacular.

I decided that if I was going to attempt a climb, I was going to do it from the bottom. Anyway, this was a week before the climbing season started and the buses weren't running. It hadn't rained in a few weeks. I was couchsurfing with a friend in Tokyo, and had no transportation to the mountain except the local suburban train. This dropped me in the town of Gotemba, at an elevation of about 1000 ft above sea level. I got there at noon, and proceeded to walk through the town the 12 or so miles to the mountain. I had just been on a 38-foot sailboat for two months, so walking felt strange. At 6pm I was at the foot of the mountain, where the volcanic scree began. I ate a full pound and a half of pasta I had brought, and promptly fell asleep for a few hours. At 9pm I woke and started up the tight switchbacks leading straight up the scree.

The lights of Tokyo lit up the clouds to the north so I could see without a headlamp.

I hiked and hiked, falling into a deep trance. The fatigue grew and grew, and I slowed down. It began to hail, but there was no shelter, only an endless slope of black scree in every direction. Just as my spirits started to sink, I came up to a hut of some kind. During the climbing season, people often stay overnight in several high-elevation huts to summit in the early morning. The hut was empty and locked, but I found some corrugated steel and huddled behind it against the building, while the hail continued. I think I fell asleep for a few hours; I was in such a state that I couldn't tell for sure.

I awoke to a strange glow on the other side of my sheet of steel. Behind the steel and the clouds, the sun was rising. With delicate awe, I watched the darkness be replaced with light. It was the most unique and the most beautiful sunrise I've ever experienced. After resting some more, I decided to head down. (Running down volcanic scree is one of the most fun things in the world, by the way.) The hail turned into rain, and stopped completely by the time I walked back to the train station.

Mount Fuji is 12,388 ft high, and I figure I made it to about 10,000 ft. I have no regrets, only awe at the beauty and power of that mountain and gratitude that I had the chance to show it my respect. Regardless of how nice it might be to reach a summit and take a cute sunrise picture, what matters is how we approach that goal, mentally and physically. It is always better not to make it than to fake it or cheat.

By saying that math is a religion, I don't mean to degrade it. Organized religion is a universal human construct, with profound value in the lives of most of us on the planet. As with all human creations, there are some potential negative aspects of religion. One is the injustice and inequality that may come from the hierarchies it creates. In Hinduism this is the caste system. In mathematics this is the knowledge hierarchy. Another is the danger of blind faith, which I discussed above in the case of mathematics.

A third danger is the development of a dogma. Shared beliefs and behaviors may become too entrenched, to the point that questioning is discouraged. The dogmas of the various world religions are not hard to recognize. Is there any dogma in mathematics?

I believe there is. I think there are quite a few unspoken, unquestioned rules among the mathematics community. Here I'll describe one example: the PhD thesis.

The official documentation I have describing the requirements for a PhD degree at my university say precisely this about the thesis: "An original contribution to knowledge. 27 credits of Math 800 required." In reality, a very specific "contribution to knowledge" is expected.

You must work for several years to generate some original mathematical results – a single significant one, or several smaller ones. Ideally, you should have one or a few papers submitted to journals for publishing. The thesis is an exposition of these results, and it is expected to have a specific form. In fact, all mathematics these days is written using the same typesetting computer program, called LaTeX. LaTeX has a thesis template. You would raise some eyebrows if you did not use the thesis template in LaTeX to format your thesis. The overall organization must follow closely to the standard form: abstract, overview/introduction, background, results, conclusion, and bibliography. The section describing your results must be written in the style of traditional mathematics: setup, theorem, proof, theorem, proof, etc, with occasional but brief guiding or motivating prose. The royal "we" must be used throughout.

Sometimes I fantasize about writing a thesis that breaks these rules. It would be in a non-traditional format, with a non-traditional organization. The mathematical content would be rigorous and (hopefully) significant, but I would present it as a journey through a mathematical space that included a human dimension. I would tell the story of my process, I would describe the experiential correlates of the results, and I would present the information in a way that put the key mathematical ideas center stage. I would weave all these components – the rigorous mathematics, the personal narrative, the metaphysical and experiential structures – so seamlessly they couldn't be separated. I would use some creative organizational framework that would be able to integrate these components holistically and illuminate their many interconnections, and do so artfully. In short, I fantasize of creating a thesis that is a glass bead game[5]. It would be more than a contribution to knowledge; it would be a work of art.

What would happen if I handed in such a thesis? I imagine being sent before a panel of old, wise mathematicians in a dimly lit seminar room with chalkboards and hard chairs, and having one say, "The results are great, but could you cut out all that extra stuff about process and humanness? That's not math." I would launch into a debate with them about where exactly mathematics stops and non-mathematics begins. I believe I could point out a few contradictions in their position.

---

[5] Hermann Hesse, *Das Glasperlenspiel*. 1943.

Math and Brahmanism

In the last essay I described how mathematics looks from the inside. Now I want to ask the question: how does the mathematical community fit into society as a whole? My answer is that mathematicians are modern-day Brahmins.

Four thousand years ago in ancient India, the religion of Brahmanism developed, as a precursor to Hinduism. It was built around sacred scriptures called the Vedas, which, among so many other things, established the caste system that has remained almost unchallenged to the present day. The Rig Veda describes how humans came from the body of the giant Purusha: the priestly caste of Brahmins came from the head, the ruling caste of kshatriyas came from the arms, the merchant caste of vaishyas came from the thighs, and the artisan/worker caste of shudras came from the feet. Below this were the dalits, also known as the Untouchables, who had no caste; they were "outcastes".

In the language of the day, "Brahma" translated as "the word", and the power of the Brahmins was centered around their ability to recite the sacred chants and hymns. The Sanskrit language was developed around this time, as a sacred language only the Brahmins could speak and understand, and it was in Sanskrit that all the sacred words were eventually written down. The sacred texts also described rituals. Recitation in this esoteric language and conducting of the sacred rituals were the only way to the gods. The people were powerless without the Brahmins and their connection to the divine.

The Brahmins would conduct their rituals to ask for good things for the people – successful harvests or business or battles, happy weddings, lots of children, etc. In exchange the Brahmins were supported, protected, and given utmost respect – the respect you would give someone who was part human and part god.

Since their power was based on this esoteric language, the Brahmins made no effort to make it intelligible to the public. Rather, they embraced the unintelligible in order to maintain their power.

It seems strange to declare oneself a modern-day Brahmin, but I think there are many analogies in the way mathematicians are seen in society. It amazes me how much respect I'm given for working in mathematics. As with the Brahmins, this respect is based on ignorance and some intimidation.

We have our esoteric language, to be sure. If you asked me to explain precisely what cohomology is, it would take hours. I would have to use other words that you wouldn't really understand either. Assuming you're not already a mathematician, it would probably take months of study before you could say you understood what cohomology is. This gives me some power over you, if you let it.

The respect given to mathematicians also comes from the fact that people think mathematics is important. Napoleon is quoted saying, "The advancement and perfection of mathematics are intimately connected with the prosperity of the state." People say that math is the language of nature or the queen of the sciences. Math is the key to nature and technology, or so we're told. The truth is that a lot of what mathematicians spend their days thinking about is so hopelessly abstract it is merely art with ideas. Maybe the science of the future will use the stable homotopy theory of today, but maybe not.

Mathematicians get respect because people remember math to be hard. But what does that mean? I think it says more about the way math is taught than about the people that can't think of any better use of their time than studying math.

In general, mathematicians encourage these beliefs. We want people to think math is hard; we want people to think math is important. We don't feel the need to explain to people that what we do is more like art than anything else. We want people to believe we have some magic connection to an Absolute realm of universal ideas that since Plato has echoed of divinity. It's not hard to see why: it gives us some of the same power and protection that the Brahmins enjoyed.

Maybe people want to believe the romantic story of the wacky mathematician, with one foot in reality and one foot in the clouds, who speaks the language of nature and can see patterns that ordinary people don't see. But I don't know how to explain the sociological emergence of the Indian caste system. I'm not prepared to claim that people like existing in a hierarchy in which others are above them.

By comparing math and Brahmanism, I hope to point out some of the unhealthy social dynamics that play out between mathematicians and non-mathematicians.

During my four trips to India, I got to see some negative aspects of the caste system there. I spent six months living in Rishikesh, a sacred pilgrimage town at the site where the Ganges River comes out of the Himalayan Mountains and onto the plains. The town is mainly a collection of yoga ashrams; this is the place the Beatles stayed during their time in India. I was working to establish a recycling program along the shores of the Ganges. The town gets many domestic tourists who come to worship the river there. The beach is cluttered with years of the plastic bags that were used for various offerings of flowers or milk.

Usha was a 26-year old woman there with three children, sleeping on the steps beside the river. She was an

Untouchable. Her husband was a drunk and abandoned her. Their single-room mud house in a distant village had collapsed in the monsoon rains. Usha had a high sing-song voice and was quick to smile. When she and her children moved through the bazaar it was a chaotic parade of distractions and squeals, like a mother duck leading her ducklings. We became friends and I visited her village to meet her husband's family and see her house. Everyone in the village was an Untouchable; they told me I was the first white person ever to visit!

Back in Rishikesh I gave Usha a job helping me clean the beach. We would pick out plastic from the sand, rinse it, and collect it for recycling. I made several dustbins for recycling, and had signs painted near the beach stating in Hindi: "This beach is your temple. You should treat it as such." I had found a reliable shopkeeper/community organizer in town who had agreed to maintain the recycling program once I left. We also set up a system for continuing Usha's employment: I would send him her salary (the minimum wage, which is a dollar a day), and he would pay her monthly; we would keep this up for as long as Usha wished to. There was enough plastic on the beach to keep her busy for a long time. (One Westerner, there to study yoga, responded to my suggestion that she help us clean the river with, "It'll take a thousand lifetimes to make the river clean!")

The next step was to get Usha a room to live in. Her oldest child was five years old, and perhaps would be able to start going to school if she wasn't sleeping outside. The only rooms available were in the ashrams. I approached the managers, all Brahmins themselves, at several yoga ashrams. Each ashram has lots of housing for pilgrims, and also cheap housing for their staff. I pleaded her case and tried very hard to get an ashram to agree to rent a room to Usha. The answer was no, and the reason, in the end, was simply that she was an Untouchable.

The problem with a caste system is the injustice and inequality that it fosters. The Brahmins may lose touch with the public, and become less able to empathize with them or in some cases even see them as human beings. The Brahmins may lose interest in going out of their way to help those below.

This comes naturally from the generations of isolation between the castes. Young Brahmins are born in the ashrams and live in the ashrams; they know little of how life works outside the ashram. Similarly, if you go straight from high school to college to graduate school, and are there taught by mathematicians who have done the same and never left academia, there is a danger of isolation and desensitization. There are many mathematicians that have never worked outside of math, and this makes it hard for them to understand the blue-collar workers that make society run. Academic elitism flourishes.

(Personally, I'm grateful for the many and diverse lessons I've learned from the many and diverse jobs I've had: lifeguard, tutor, construction worker, farmer, mason, lab assistant, deckhand, apple picker, election worker, math TA and teacher.)

In India I sat for religious discourses given by holy men. Devotees would crowd around their Teacher, gathering the pearls of wisdom falling from his mouth. Many of those teachers allow themselves to be seduced by the power of the situation; there are more than a few accounts of teachers abusing their power and seeking fame, sex, or money.

I watched the old, wise holy men wandering around the town and ashrams. They were treated with utmost respect, although mostly all they did was sleep (a lot), eat, walk, and meditate. In the math department here there are definitely those old wise men. They don't seem to teach or do research any more, but they are on the payroll, supported out of respect for their wisdom I guess.

The Brahmins do depend on the public for support, and this can be unhealthy some times. In Rishikesh I had a close Brahmin friend named Panchannan. He had the charm and looks of a movie star, with a naïve gentleness and curiosity. Panchannan would perform rites and conduct various ceremonies for money, but he was worried that he wasn't making enough to support his mother and father and siblings. He was very worried that he might have to get a job outside the ashram.

To avoid this problem in mathematics, the mathematicians have convinced the public that people must learn math, and people must be taught math. Neither is really true. There's not much of a reason for everyone to know calculus, for example, or to be able to do things that calculators can do better.

In the summer of 2008 I taught my own section of Math 111, Business Math. In the first week of classes I told my students that they didn't have to come to lectures if they didn't want to, that I was mostly just rephrasing the textbook in my own words, and that they could teach themselves. In fact, I think the most important thing you can teach someone is that they can teach themselves. To the old saying, "Give a man a fish and you feed him for a day. Teach a woman to fish and you feed her for a lifetime," I'd like to add something like, "Teach someone that they can teach themselves, and you give them anything they could ever want."

When I told my faculty mentor that I was encouraging my students to teach themselves, he said something to the effect of, "Yeah, I don't tell them that because I don't want to lose my job." When I discussed it with another professor, he said, "I know, it's all smoke and mirrors anyway, isn't it?"

Another danger in isolating the priest caste is inbreeding. Seriously. I remember going to the incredible Nataraja temple in Chidambaram, in South India. This is the place where the Dancing Shiva is said to have danced his/her dance of creation and destruction. It is also where they keep the urn that is said to contain the ether of consciousness. The temple is huge, and sustains a significant population of priests. These Brahmins are known for their strange hair cuts and crossed eyes. The crossed eyes come from a lack of genetic diversity among the small community, developed over centuries of inbreeding.

Lack of diversity is also a serious danger to mathematics today. It leads to stagnation, delusion, and sickness, and I think this can be seen in some areas of math. I'll discuss this lack of diversity more in the next essay.

Mathematicians are not Brahmins. We are artists. We deserve as much or as little respect and support as are afforded all other artists.

Math and the Math Culture

Now I want to look at the math culture from the outside. I believe that math is a human endeavor; the body of knowledge we call mathematics and the community that does mathematics have been co-dependently evolving for millennia. I'd like to try to describe prominent characteristics of the math culture and suggest how aspects of math itself may have helped to bring about these characteristics. Conversely, I'd also like to suggest ways that our methods of doing math have affected the development of mathematical knowledge.

The math community is known for a level of absent-mindedness and eccentricity. I discussed this phenomenon in the essay *Math, Hitchhiking, and Couchsurfing: Malfunctioning in Society*, along with possible origins in and implications for the nature of mathematics itself.

Mathematicians focus on those things that are objectively true, that can be proved. This is a common theme in these essays: the math culture is obsessed with objectivity and the proofs they think of as demonstrating absolute truths. As a result all the lectures, teaching, and writing of mathematicians are proof-centered. This is a cultural phenomenon. It doesn't have to be this way. Some people have suggested that the central element of mathematics, the thing that makes something math, is not the proof but the idea[6]. More fundamental than the logic constructions are the nebulous things they interconnect – the ideas. Ideas have an objective component in the symbolism

---

[6] For example, see *How Mathematicians Think*, by William Byers, 2007.

and rigor and a subjective component in the personal experience of "getting it". By centering mathematics on the ideas that are being created and communicated, we can take a step towards a more coherent and holistic understanding of math and the mathematical experience.

But mathematicians in general don't think directly about the ideas. Your job is to write rigorous proofs, not to describe your intuition of "what's going on". We're taught that the more objective and impersonal the proof is, the better it is. The reader is left to personalize the content, and to independently discover the ideas underlying the proof. Sometimes it seems the author himself or herself didn't quite grasp the idea of "what's really going on". Indeed, figuring this out is the hardest part.

As a result of this attachment to the universal objectivity of math, all the intuitive, personal dimensions are devalued and swept under the rug.

Why does the math community approach math in this way? I don't know. Historically, this has always been the case. From the time of Pythagoras through the Renaissance and up to the 20th century, math has been closely associated with theology and the Absolute. Some might even define mathematics to be those mental constructions that can be objectively described and manipulated.

Maybe the simplest answer is that math just feels so objective. In so many ways, when working with math, ideas take on an objective quality. It's very easy to feel like one is exploring a Platonic realm and discovering universal truths. There is an excellent account of how this objectivity might emerge from the social construction that is mathematics in the book *Where Mathematics Comes From* by George Lakoff and Rafael Núñez.

Learning mathematics is centered around solving problems. You don't learn math by just reading, or just listening to someone lecture. Every math course consists of some lectures or reading assignments and lots and lots of homework problems. There are frequent quizzes and exams that test your ability to solve problems. In math, if you can solve problems, then you understand the material. In my experience, the emphasis on problem-solving is greater in math than in other disciplines. I think this is because math is easily presented in an objective way, but any understanding of math is a personal, intimate understanding because math is so imaginary. On the one hand, math is precise and universal; we can talk about it clearly and efficiently. On the other hand, these impersonal ideas must be personalized to be understood, and this personalization takes place deep inside each student's own mind. A teacher can show you the door, but you must walk through it. Since mathematicians have made almost no effort to understand this intimate internal formulation of math, we've had to base our teaching on the one technique that has proven successful: solving problems. When presented with a problem to solve, a student must personalize the ideas and actively traverse his or her internal math world to arrive at a solution. What's more, once a personal understanding is reached, the solution must again be impersonalized and written down for others to examine. In this way, problem-solving is well suited for learning to navigate between the objective and subjective dimensions of the math experience.

How has this emphasis on problem-solving in turn shaped the development of mathematics? As I said earlier, mathematicians usually proceed by asking, "What can we do?" rather than "What should we do?" Want to advance math? Here's the recipe we're taught. You must find a good problem. This problem must be tractable and not too hard, but also interesting and not too easy. Spend all your time trying to solve this problem. Whether or not you succeed, new problems will suggest themselves. Continue this process as long as you can.

As a result, various areas of mathematical study are continually clarified and enriched. What sometimes feels lacking, however, is the large-scale vision that has the potential to revolutionize our understanding. If you only ask questions you think you can answer, you'll never know what you don't know.

At the higher levels, mathematical knowledge is transmitted through a master-apprentice relationship. At the lower levels, the teacher-student relationship suffices. The transition occurs at the same point that the learning changes from digestion to production. To learn for the sake of knowing math, classrooms and homework are good enough. To learn for the sake of understanding math deeply – living and breathing math – we must apprentice under a master. I'll choose a field of mathematics to specialize in and select an advisor that does this work. I'll immerse myself in this subfield while surrendering to my advisor's wisdom and experience. I'll learn more than math content; I'll learn how to live like a mathematician, how to look and act like a mathematician, how to create mathematics. To a certain degree, this is learned through mere imitation.

All doctorate programs use this advisor-advisee format. So do most fine arts and skilled trades. Throughout the centuries, the way to become a carpenter, blacksmith, musician, or scholar has been to find a master and follow in their footsteps.

I think the fact of this established tradition says something about the nature of mathematics. At its lower levels, it is functional and egalitarian. At its higher levels, it is an art. To be a mathematician is more than to be a teacher or a producer of papers (or, as the Hungarian Alfréd Rényi said, "a machine for turning coffee into theorems"). We must learn a way of life. What does this mean? The math community has established some social norms; to be part of the community and work and create within that community, you must learn these norms. We must learn the art of mathematics. I don't think anyone knows how to teach this art, or any art really. All we know is that by apprenticing under a master, you might figure it out on your own.

What has been the effect of the master-apprentice relationship on the development of mathematics? Throughout the history of mathematics, math has been localized around various schools of study and schools of thought. The masters of mathematics are spread throughout the world. Surrounding a given master grows a small community of apprentices and devotees. These groups share more than just a common field of study, they share a common way of thought derived from the personality of the master. In this way, the various subfields of math each develop their own flavor – a particular philosophy or approach to doing mathematics.

So we had the competition between the English mathematicians adhering to Newton's formulation of calculus and the French mathematicians adhering to Leibniz's formulation of calculus. We had the Bourbaki group and their foundationalist approach to vast areas of algebra. Although mathematicians strive to remove the personal dimension from math and present it in a universal form, all of math is historically embedded. It is debatable whether the localized social norms can ever be extricated from their respective fields of study.

I've commented in several places about the intensity of the first few years of math grad school. You've probably realized that those were tough times for me. During this time, quantity is more important than quality. You must demonstrate your endurance, and only after you've done this will you be allowed to demonstrate your creativity.

Why is it this way? I'm sure it's because doing research on the frontier of math is really nice, but you have to get there first. There's a lot of ground to cover in order to get to the frontier, and it's in the interest of the university and math community to get you there as soon as possible. Before you can specialize and head to the fringes of math, though, you must have a strong understanding of a significant core of mathematics. Most mathematicians will need to teach math in order to support themselves, and for this you must have a reliable toolbox of math knowledge from which to work. Even if you know you want to devote your life to researching partial differential equations, in order to get paid you may need to teach algebra or topology at an undergraduate or graduate level.

I think mathematicians undervalue creativity, but they know the value of endurance. It is not uncommon to spend months and years working on a problem in order to extract some result. Sometimes the problem proves impossible, or someone beats you to the answer. Mathematicians must become accustomed to working productively in spite of this uncertainty and frustration.

Maybe it's because mathematicians never study *how* we do and learn math, so they don't know how we might do and learn math best. The best idea they have is to just do a lot of it.

There are definitely consequences for the evolution of mathematics as a whole. Most of the creative, non-traditional grad school peers of mine couldn't handle the stress and all-encompassing intensity of those first years, and they quit or left with a Masters degree. It's clear to me that math has suffered as a result. Many new and potentially revolutionary ideas that these people carried will not make it into the collective mind of the math community.

Taking this idea further, one thing that often strikes someone who is looking at the math community from the outside is the lack of diversity. There is a very specific type of person that succeeds in the math institution of today, with remarkably specific types of behaviors and way of life. Presented with this narrow mold, many, many people are weeded out: the creative ones, the ones with diverse interests, the ones with irregular work habits, the ones with chaotic lives. This lack of diversity in lifestyle and life experience leads to a lack of diversity in thought. The mold narrows, and the stereotype propagates itself. There is more of the same; the status quo is maintained.

Have you ever walked in a forest, or visited an organic farm? Aren't we all taught the immeasurable value of diversity – diversity in thought, in lifestyle, in environment, in experience? When there is diversity, new things emerge – things that couldn't be foreseen by the wise elders. When there's diversity, relativism establishes a philosophical chaos that frees us from the prisons of thought and action we have created for ourselves without even realizing. Deterministic control and power hierarchies are dissolved and cooperative, collective evolution replaces it. Diversity is not valued in the math community as much as it should be.

Why might this be the case? Math has an incredibly objective feel to it. When you know some math, you tend to think you're right. When someone else doesn't know that math, it seems clear how to proceed: you must lead them to the place where you are, so they can look at the landscape from your perspective. You know the answer to the questions already. If someone wishes to disagree or diverge from the path you followed, they should be discouraged and taught the truth as you know it.

All the older mathematicians have focused their lives on smaller, more specialized arenas of mathematical truth, because everyone knows it's now impossible to know all the types of math. Many have little patience for stepping out of their math world into another's for very long. Some seem to be thinking, "Why should I try hard to understand your strange or seemingly naïve type of math, over there? I've already committed to my type of math, over here."

What's more, it seems that most mathematicians don't have a very clear sense of what exactly math is or a very long-distance view of where it's headed. What they do know is what math has been, and where it came from. As a result, there is a lot of looking back at how things were done. Math is old and robust; for the most part it proceeds in very small, careful steps. I'm sure this explains some of the conservatism in the math community.

This conservatism and lack of diversity propagate themselves. Math continues to grow in small, careful steps. (See the final essay, *The Next Math Revolution,* for a discussion of the exceptions to this.) The math community maintains its narrow mold and specific social norms. The threat of stagnation and delusion remains.

Sometimes when I listen to the radio in America, or go to live shows, I feel disappointed by a lack of diversity. Music is particularly transparent – if there is a strong creative impulse in the music it jumps out and ropes you in, and if there isn't then the music leaves you uninspired. America was built on diversity, but sometimes I wonder if some cultural arrogance is keeping us turned too much inward.

This past winter when I traveled in the Middle East and the Balkans, I went to many clubs and live music shows. There you find brilliant artists that are combining old and new, local and global sounds. A veritable dance and music revolution is taking place, and while more and more American musicians are starting to notice, I think many Americans are still oblivious. I've never danced with such joy as I did in the clubs of Tel Aviv, Beirut, Istanbul, and Belgrade. The Middle Eastern melodies, the funky Turkish and Eastern European rhythms, the Balkan brass tradition, the Gypsy enthusiasm and spontaneity – they each can inject into and synthesize with global electronic, dance, rock, and punk sounds to create magic.

I hope to join a band here in Seattle that will be a part of that movement bringing the Balkan sound into American music. This brings me to my next observation about the culture of mathematics: varied lifestyles are discouraged. Mathematicians are supposed to be singularly dedicated to their research. Ideally, they would live somewhat monastic lives, removed from the distractions of society. We're taught that it's okay to have a family and a hobby or two, but to be truly successful a mathematician must have one and only one passion: mathematics.

What about math itself might have brought about this cultural phenomenon? Well, math is challenging and there are a lot of smart mathematicians. To keep up with the current research, and to stay in top form, a focused lifestyle is helpful.

In medicine or law or theater it might be encouraged to take time off after getting a degree in order to get some diverse life experiences that will later inform your work. However, many mathematicians don't recognize that diverse life experiences might inform mathematics; because all they've ever done is math, they don't know. What's more, math is easily forgotten, and if you're not continually working with it you will lose it or get rusty.

Besides having a focused lifestyle to stay immersed in math, it is very useful to have a simple life. This is because of the delicate complexity of the structures of math, which for most people require a clear and quiet mind to be understood. If you're filling your time with many things, or investing passionately in other endeavors, then in the time that you are working on math your mind will often be unsettled and loud. Thus it's natural for the math community to sanction and encourage a simple, almost monastic lifestyle with few distractions.

I think another reason for this is that math is invisible. It is very hard to gauge your progress in math, at least until you are well-established (in which case you are certainly free to do whatever you want with your time). Before then, the only way of showing you're passionate and committed to mathematics is by doing a lot of it. How do I justify spending a Friday night staying in my room doing math for hours and hours, especially if it is reading or research as opposed to some homework problem? While my friends are out undertaking various adventures, I'm busy advancing my mathematical knowledge a few small steps. I have almost nothing to show for these many hours of intellectual struggle. No matter how much I enjoy doing math, I also enjoy other Friday night adventures. One convincing justification is this story of the mathematician: mathematicians aren't supposed to have varied lives, they're supposed to be simple and boring, undistracted by the many non-mathematical adventures in life.

One really fun way to tease a mathematician is to ask them, after a weekend or vacation, what it was they did with their free time. They'll scoff at the suggestion that they were goofing off, and insist they were working hard at math, maybe even attempting to explain the progress they made.

Math is invisible and we often can't understand each other's different specializations. Every human looks for occasional validation. The only way a mathematician can find validation is often in garnering respect for their dedication and hard work.

I think this aspect of mathematics – that it is so invisible and hard to explicate – propagates itself unnecessarily. Seeking validation and respect, mathematicians embrace the unintelligible. As I discussed in *Math and Brahmanism*, by endorsing the idea that math is hard and mathematicians speak a magical language the public cannot understand, we are given respect and support. This respect and financial support is undoubtedly enjoyed. Higher mathematics is removed from public consciousness and stays that way. Mathematicians isolate themselves, become removed from society, and stay that way. The math they create continues to float in the clouds, detached from earth, and remains unintelligible to the non-mathematicians. We keep math hard and don't try very hard to explain it to non-mathematicians.

This is a cynical interpretation blaming mathematicians as being power-hungry, lazy, and selfishly manipulative, but there is a gentler one. It is true that to explain higher math concepts to non-mathematicians – so they understand as a mathematician does – would take a very, very long time. There is a specificity and precision inherent in mathematics and demanded of mathematical discourse within the mathematical community. When a mathematician tells a non-mathematician, "You can't understand", he or she is making the simple mistake of thinking that to understand means "to understand as I understand". There are many different ways of understanding, if we step outside of the specificity and precision that is required within the math community. I have occasionally succeeded in explaining math to my friends or family, in a way that they "understand" math in a non-mathematical way.

The implication of this is that perhaps, if we can overcome our fear of losing our jobs, we can step outside of mathematics and deliver to the people some of the beautiful mathematical secrets we've been keeping. We can tell them about Cantor's infinite hierarchy of infinities, each infinitely larger than the previous. We can tell them the story of non-Euclidean geometries. We can tell them about the Continuum Hypothesis and the unknowable mysteries it implies are held within a simple continuous line. We can tell them about Gödel's Incompleteness Theorem, and its implications for the pursuit of knowledge.

Part IV: Partitions of Unity

Math and Spirituality

One of the most beautiful and well-known icons of India is the bronze statue of the Nataraja, or the Dancing Shiva. Shiva, in a half-male, half-female form, is dancing within a circle of bronze surrounded by flames. In one hand is a drum and in the other is a flame. One foot is suspended in motion and the other is standing on a small demon.

The usual interpretation is that the demon is ignorance, the drum is creation, and the flame is destruction. The circle in which the dance takes place is reality. The Nataraja is dancing the dance of creation and destruction in order to sustain reality and to overcome our ignorance.

One way to understand reality is using the concepts of creation, destruction, and the interplay between the two. Another way is good vs. evil, or order vs. chaos, or yin vs. yang. Each of these is a model for reality, and no model is perfect. Drums can destroy and fires can create.

It's not uncommon for people to confuse models or symbols with the things they are meant to represent. The word is not the thing. It can be quite frustrating to try to describe the indescribable.

Yet we don't throw out our models, no matter how flawed they might be. We can't. They are all we have to try and understand. Without any models or any language, all we have is ignorance. This is the human condition. There is some perfect Reality we cannot access directly. We can only make models and frameworks, and project that Reality down into our imperfect models. Each model is its own universe, and expands to create a space in which we can observe the shadows of Reality. The dance of the Nataraja is restrained to within the circle, and using the dance we cannot step outside and see the entirety of Reality. But that dance also sustains the space of the circle, in which we can watch the movements of creation and destruction. Without the dance, the circle would collapse into ignorance.

I believe our understanding is always limited, and there is always something outside of any language or model we might create. This is a belief: that there is more to things than meets the eye. I think this is the origin of spirituality. I think a spiritual life is one that strives for the unattainable goal of experiencing the Whole.

There's a small tattoo on my right wrist – a square inscribed in a circle. It's meaning to me has continually changed as I've grown. I originally got it when I was 18, while working at an organic farm for a summer, and it was inspired by that experience and the Hermann Hesse novel *Narcissus and Goldmund*. Today it reminds me of the human condition – the square represents the frameworks and languages that we create and must live within, and the circle represents the transcendent Truth or Perfection that we are striving for but will never attain.

There are also many funny stories surrounding my tattoo. I remember one night sitting on the beach in Goa watching the sun set on the ocean. Another traveler – an Austrian surfer – came over and we started talking. When he saw my tattoo he became quiet. He locked eyes with me in an intense stare and said, "Whoa, man. Do you go to the meetings?" I stared back at him, was quiet for a little while, and with a slow nod of my head responded, "Yeah, man." We kept the stare for a bit longer, and then without another word went back to watching the sun set. I never found out what meetings he might've been talking about.

Mathematics might be defined as the creation and exploration of conceptual frameworks. Every field and subfield of math is itself a small universe, framed by various definitions and axioms, in which mathematical objects can be created, manipulated, and studied.

It is often the case in mathematics that we define an object, or even prove it exists, but then struggle to comprehend it. In order to develop an understanding, we will project this object down into another, usually "smaller", framework in which we can better see it. We may attempt to construct simple examples and study their properties, in order to develop some intuition about what is going on in the general case.

On the other hand, given any mathematical framework there is always the drive to transcend it and generalize. While working within a system, you will every so often get a glimpse of some larger truth. This glimpse may reorganize ideas to reveal deeper structure or insights within the system, or it may suggest a way to break out of the system and expand.

The experience of this glimpse of a deeper understanding of "what's really going on" has many times been described using language similar to that used to describe spiritual realizations or awakenings. Mathematicians are continually enlarging their models to contain more and more generality and more and more complexity. This is a spiritual pursuit.

The Next Math Revolution

The invention of zero caused a mathematical revolution. It took a long time for early mathematicians to get used to this seemingly impossible idea – the thing that is nothing. But once it had been accepted into math, it changed the notion of 'number'. A number did not necessarily have to correlate with a collection of things, because zero is a number that correlates with no collection. A number could be freed from the physical world; ideas could be motivated by the physical world but were not bound to it. Soon the negative numbers were invented, then the rational numbers. This drive towards abstraction permeates all mathematics now.

Incidentally, it is thought that zero was invented independently around 300 BCE in India and by the Maya. One explanation for the Indian genesis is that in ancient Indian philosophy the idea of Nothing figures very prominently. The Indians were quite comfortable with the idea of nothing being something.

The idea of infinity has always been a challenge for mathematicians. The type of infinity implied by the ellipsis in "1,2,3,..." is called potential infinity. It can be proved that $1/2 + 1/4 + 1/8 + 1/16 + ... = 1$. (Imagine taking a string one foot long, cutting this in half, cutting one of the shorter pieces in half, cutting one of the shorter pieces in half, and so on.) This second type of infinity is called actual infinity; by writing this equation we are equating the object "1" with the result of an infinite process. Thousands of years ago the Greeks were already fine with potential infinity, but couldn't accept the idea of actual infinity. As a result, while the Greek mathematicians were prolific in the field of geometry, their study of abstract symbol manipulation stagnated. It was the Arab mathematicians – who for some reason had less trouble with actual infinity – that invented and established algebra as part of mathematics.

The question of infinity was not quite settled though. In the 19th century, mathematicians were trying to understand the relationship between the discrete and the continuous. The German mathematician Georg Cantor single-handedly revolutionized math when he formulated an elegant and comprehensive mathematical theory of infinity. Cantor proved that there are different sizes of infinity. In fact, there is an infinite hierarchy of infinities in which each infinity is infinitely larger than the previous. For example, the infinity of points on the real line is "larger" than the infinity of points in the list 1,2,3,... However, the infinity of points on the real line is the same size as the infinity of points between zero and one on that line!

Many mathematicians of the time reacted vehemently. Cantor was called a "charlatan", a "renegade", and a "corrupter of youth" for teaching his theory. Poincare called the theory a "grave disease" infecting mathematics. Theologians accused him of questioning the absolute omnipotence of God. (Cantor, on the other hand, believed this theory had been communicated to him from God.) Cantor didn't respond well to the criticism, and succumbed to depression. From 1884 when he published his results until his death in 1918, he was continually in and out of sanatoria. It's argued that the mathematical community was in part responsible for this mental illness.

Today, of course, Cantor is considered a genius. His theory of so-called "transfinite numbers" revolutionized mathematics, and has had far-reaching consequences.

A third example of a revolutionary idea is randomness. It is not uncommon for math to be described as the study of pattern.

By definition then, randomness cannot be integrated into mathematics. But we've done exactly that, in probability theory. Mathematicians have brought their rigor and precision to study the pattern that is no pattern. Now probability theory is a rapidly growing field with many powerful applications.

What is the recipe for a mathematical revolution? Start with some strange idea that is not math (as we know it), and could not possibly be math (as we know it). Something like nothing, or infinity, or randomness. (Math is rigorous and precise, and studies definite objects – it can't study "nothing" or "infinity"! Math is the study of pattern – it can't study "patternlessness"!) The only way math can come to include this strange new idea is by growing, and because the idea is so strange, that growth will be dramatic. At the heart of the contradiction, find the revolutionary breakthrough that will change what we think mathematics to be. Math may lose something – some smaller way of thinking or attachment to a historical attribute – but it will gain rich new ideas and wider applicability and generality.

After the breakthrough, mathematics will mean something different, and the new strange idea will prove to be a powerful and deep source of new mathematics. Over time, it will be interwoven with many older ideas, and then it'll seem to pop up everywhere. The possibility that it might *not* be a part of mathematics will be almost inconceivable.

So then, what is the next math revolution? What is the strange new idea that we must integrate into mathematics? Subjectivity.

Subjectivity could not possibly be a part of mathematics as we know it now. Mathematics is objective, absolute, and independent of personal experience, right? Mathematical objects are those mental constructions that sustain a structure independent of the personal whims of any given mathematician. Mathematical properties are those that can be described precisely and understood by any and every mind that takes the time to do so. The development of math has been a continual journey towards more rigor, more precision, more specificity, and more objectivity. How could subjectivity possibly be integrated into mathematics?

I think that it is exactly because math has become so rigorous, precise, and "objective" that we're now prepared to break through to something new.

In the late 19th century, geometers were shocked by the discovery of non-Euclidean geometries; the fifth axiom of Euclid was unnecessary. At the same time, analysts that had been struggling since the invention of calculus with questions surrounding the discrete and the continuous were finally able to answer these questions by improving mathematical rigor. Thus math entered the 20th century with these convictions: we must establish mathematics on secure foundations, and we must bring the utmost rigor and precision to mathematics. With more rigor and precision, math seemed to become even more objective and universal, and as we probed deeper into the foundations of mathematics the philosophical debate began: Platonism or Formalism? More than ever, mathematics seemed to be a visit to Plato's cave of Ideas. At the same time, more than ever, mathematics seemed to be an impersonal manipulation of logic, a meaningless game of symbols.

These two philosophies were obsessed with objectivity and left no place for the subjective mathematical experience. Even at that time, some mathematicians realized how unsatisfying these two philosophies were, and how unfaithful they were to the actual experience of doing mathematics. Brouwer and the Intuitionists suggested an alternative and recognized that mathematics should in some way take into account the mathematicians, but their insistence on constructive proofs doomed them to be marginalized[7].

Once Gödel published his Incompleteness Theorem, proving that in any sufficiently complex mathematical system there will always be statements whose truth can be neither demonstrated nor disproved, the foundations movement was abandoned. As postmodernism spread through academia, some new approaches were suggested to move beyond the Platonism/Formalism stalemate. What if mathematics was all just a social construction? How then to explain the properties it seems to maintain – its consistency, universality, and objectivity?

In order to understand these questions, it became necessary to look at the mathematicians themselves and see what exactly it means to do mathematics. This process has just begun recently, in the last twenty years. In philosophical circles, there is some discussion of a humanist philosophy of math based on actual mathematical experience rather than ideal *a priori* notions. This discussion hasn't quite made it into the community of mathematicians, though; most mathematicians are still living with a contradictory mix of Platonism and Formalism. But history has set the stage for a mathematical revolution.

Objectivity has come to be seen as central to mathematics, but in fact it is only one aspect. As William Byers suggests in his book, *How Mathematicians Think*, we must shift to center mathematics on the ideas themselves. It is the ideas that make math what it is. It is the ideas that carry meaning; they are

---

[7] A constructive proof is one that avoids using the law of the excluded middle, which states that either a statement or its negation is true. In particular, a proof by contradiction – showing something is true by showing it's not not true – is not a constructive proof.

fundamentally what we are striving to create, communicate, understand, teach, and learn. These ideas have an objective dimension, in their precise formulation and symbolism. They also have a subjective dimension. Ideas don't exist anywhere but in our minds, as abstract patterns within a network of other ideas that we can visit with our awareness. Awareness of an idea is experiential – the idea looks like something, sounds like something, has a spatial component, etc. The experience of that idea depends immensely on the mind that contains it. Even the same idea in the mind of the same person will change and evolve over time.

Ideas are both objective and subjective, and I think by exploring this duality we can find the way to integrate subjectivity fully into mathematics. There is a middle ground between the two poles. A room of mathematicians will all experience a given idea differently, but they can agree on certain attributes of that idea, *and* on certain attributes of the experience of thinking that idea. This is called intersubjectivity – when a group of people can give a degree of objectivity to an experiential object by together communicating and sharing their own internal experiences.

I think that mathematicians can find some common ground in their mathematical experiences. I think that if mathematicians become more aware of their process and discuss their experiences with each other, we will see that math ideas have an intersubjective reality to them. For example, the intimate "Eureka" moment attached to a deep mathematical realization has been described by many mathematicians, and the descriptions are all remarkably similar. The "Eureka" moment, although it is personal and subjective, seems to be a sort of objective and universal mathematical experience.

The objectivity in mathematics that is insisted upon today is unnecessarily restricting. We can bring subjectivity into math by changing our definition of objectivity to include some intersubjective dimensions. This is the breakthrough.

This may seem crazy. But there is a 2,500-year old example of this subjective objectivity being used to build a rich, consistent, and profound body of knowledge: Buddhism and the study of consciousness.

The study of consciousness has presented many challenges to Western science. Scientific materialism – the philosophical view that all phenomena are ultimately founded on the workings of a material world – has become the dogma of science. Adherence to this philosophical perspective has made it very difficult to scientifically study consciousness, the most subjective and immaterial of all phenomena. What would it even mean to have a science of consciousness? The naïve approach suggested by scientific materialism is to reduce all the complexity of consciousness to brain chemicals and patterns of neuron firings. We don't understand how, but surely the unfathomable complexity of our neurobiology must be the source of the unfathomable complexity of sentience, right? Much progress has been made in this direction, and many physical correlates of mental experiences have been mapped out. The hard problem remains, however: how can we describe, in a satisfying way, the subjective merely in terms of the objective?

Without the dogma of scientific materialism, Buddhist contemplatives have been able to develop a much richer, more precise, and more natural science of consciousness. It is still accurate to use the word "science" because this Buddhist theory of mind has been developed using little more than the scientific method as it is known in the West.

Buddhist contemplatives conduct experiments on their mind, using their mind. In order to conduct an experiment, you must use powerful and precise equipment. As our minds in their untrained state have a tendency towards agitation and dullness, many meditation techniques have been developed to strengthen the light of awareness. It is sharpened, focused, and made more stable and reliable. Then this powerful awareness is used as a tool to probe the mind and conscious experience.

These internal experiments proceed as in a laboratory. Begin with a certain setup. Use the tool of awareness to examine some mental phenomenon or manipulate some mental object. Observe. Conclude the experiment (i.e. stop meditating), and examine the results. Create models to understand your findings, and hypothesize relationships governing what's going on. Develop a new experiment to test your hypotheses, and repeat the process.

If this were all that the Buddhists did, then the best we could hope for would be thousands of independent sciences of mind, one for each meditator. But just as Western science insists on repeatability of experiments and is interested in developing universal laws of nature, the Buddhists were interested in coming to understand those mental phenomena that were shared among all conscious beings. So after each individual conducted his or her own mental experiment, this was followed up with conversations between individuals. These conversations identified and focused on the experiences and mental dynamics that were common among minds. This is the meaning of intersubjectivity – by collectively analyzing our personal experiences, we can develop a common shared framework for understanding those subjective events.

A meditator may say, "Sit quietly and do X with your mind. See what you find. I found Y." Anyone with a sufficiently trained awareness is invited to conduct the experiment, and, according to Buddhist writings, the results are consistent. More often than not, the experimenter will also find Y. These experiments are repeatable, and their results demonstrate dynamics of consciousness that in a way transcend the individual mind. This is a science of consciousness.

In this sense, objectivity in mathematics is nothing more than a lot of intersubjectivity. Math seems so objective because two mathematicians, starting from the same set of axioms and following the same logic steps, will arrive at the same conclusion. In fact, a proof is just a description for a mental experiment. "I started here and did such and such, and arrived at X. I've retraced my steps, and presented them succinctly in this proof. Now I invite you to start from the same place, and follow my steps, and see if you also arrive at X." This is what mathematicians mean when they say math is objective. When we sit down together and talk about our journeys through our math worlds, we agree about some things.

Up until now, however, when describing our mathematical journeys we only discuss the objects we saw and the places we reached. If we widened the discourse to include the experience of those objects and the process of the journey, then I believe we'd still agree about many things. The intersubjective reality we could create through such a holistic approach would enfold all the beauty and profundity of the mathematical experience into mathematics itself.

Bibliography

Bagchee, Sandeep. *Understanding Raga Music*. Eeshwar, 1998.

Byers, William. *How Mathematicians Think*. Princeton University Press, 2007.

Capra, Fritjof. *The Tao of Physics*. Shambala, 1975.

——. *The Turning Point*. Shambala, 1982.

Dass, Ram and Paul Gorman. *How Can I Help?* Knopf, 1985.

Davis, Philip J. and Reuben Hersh. *The Mathematical Experience*. Mariner Books, 1981.

Davis, Philip J. *Mathematics and Common Sense*. A K Peters, 2006.

Enzensberger, Hans Magnus. *Drawbridge Up: Mathematics – A Cultural Anathema*. A K Peters, 1999.

Ernest, Paul. *Social Constructivism as a Philosophy of Mathematics*. SUNY Press, 1998.

Gleick, James. *Chaos*. Penguin Books, 1987.

Hersh, Reuben. *What is Mathematics, Really?* Oxford University Press, 1997.

Hersh, Reuben, ed. *Eighteen Unconventional Essays on the Nature of Mathematics*. Springer, 2006.

Hesse, Hermann. *Narcissus and Goldmund*. Bantam Books, 1930.

——. *Das Glasperlenspiel*. Bantam Books, 1943.

Iyengar, B.K.S. *Light on Yoga*. Schocken, 1976.

Lakoff, George and Rafael Núñez. *Where Mathematics Comes From*. Basic Books, 2000.

Macy, Joanna. *World as Lover, World as Self.* Parallax Press, 1991.

Rancière, Jacques. *The Ignorant Schoolmaster*. Stanford University Press, 1991.

Schoenfeld, Alan H. *Mathematical Problem Solving*. Academic Press, 1985.

Tymoczko, Thomas, ed. *New Directions in the Philosophy of Mathematics*. Princeton University Press, 1998.

Varela, Francisco and Jonathan Shear, ed. *The View From Within: First-person Approaches to the Study of Consciousness*. Imprint Academic, 1999.

Wallace, Alan B. *Contemplative Science: Where Buddhism and Neuroscience Converge*. Columbia University Press, 2007.

www.ingramcontent.com/pod-product-compliance
Lightning Source LLC
Chambersburg PA
CBHW030800180526
45163CB00003B/1105